The Laboratory
BIRD

The **Laboratory Animal Pocket Reference** Series

Series Editor
Mark A. Suckow, D.V.M.
Freimann Life Science Center
University of Notre Dame
South Bend, Indiana

Published Titles

The Laboratory Bird
Critical Care Management for Laboratory Mice and Rats
The Laboratory Canine
The Laboratory Cat
The Laboratory Ferret
The Laboratory Guinea Pig, Second Edition
The Laboratory Hamster and Gerbil
The Laboratory Mouse, Second Edition
The Laboratory Nonhuman Primate
The Laboratory Rabbit, Second Edition
The Laboratory Rat, Second Edition
The Laboratory Small Ruminant
The Laboratory Swine, Second Edition
The Laboratory *Xenopus sp.*
The Laboratory Zebrafish

A Volume in The Laboratory Animal Pocket Reference Series

The Laboratory
BIRD

Douglas K. Taylor

Division of Animal Resources,
Emory University, Atlanta, Georgia U.S.A.

Vanessa K. Lee

Division of Animal Resources,
Emory University, Atlanta, Georgia U.S.A.

Karen R. Strait

Division of Animal Resources,
Emory University, Atlanta, Georgia U.S.A.

CRC Press
Taylor & Francis Group
Boca Raton   London   New York

CRC Press is an imprint of the
Taylor & Francis Group, an **informa** business

CRC Press
Taylor & Francis Group
6000 Broken Sound Parkway NW, Suite 300
Boca Raton, FL 33487-2742

© 2016 by Taylor & Francis Group, LLC
CRC Press is an imprint of Taylor & Francis Group, an Informa business

No claim to original U.S. Government works

Printed on acid-free paper
Version Date: 20150730

International Standard Book Number-13: 978-1-4665-9362-6 (Paperback)

Library of Congress Cataloging-in-Publication Data

Taylor, Douglas K., 1969- , author.
  The laboratory bird / Douglas K. Taylor, Vanessa K. Lee, and Karen R. Strait.
  p. ; cm. -- (The laboratory animal pocket reference series)
  Includes bibliographical references and index.
  ISBN 978-1-4665-9362-6 (alk. paper)
  I. Lee, Vanessa K., 1977- , author. II. Strait, Karen R., 1978- , author. III. Title. IV. Series: Laboratory animal pocket reference series.
  [DNLM: 1. Animals, Laboratory--Handbooks. 2. Birds--Handbooks. QY 39]

  SF406
  636.088'5--dc23
                                                                    2015029265

Visit the Taylor & Francis Web site at
http://www.taylorandfrancis.com

and the CRC Press Web site at

contents

preface

There are several avian species that have been useful in studies of disease. Although not used nearly as commonly as mice, they are nonetheless important models in some areas of study. Chickens have made the most significant contributions historically as they were instrumental in the characterization of retroviruses, oncogenes, and lymphocyte cell lineages. They continue to be important models of several disease conditions, including muscular dystrophy and infectious diseases. In addition to chickens, the Japanese quail (*Coturnix japonica*), mallard duck (*Anas platyrhynchos*), pigeons (*Columba livia*), and several passerine species are all commonly used today for studies in a wide array of disciplines from environmental toxicology to behavioral neuroscience. The publication of the chicken, zebrafinch (*Taeniopygia guttata*), and pigeon genomes in 2004, 2010, and 2013, respectively, only adds tools to the research armamentarium and makes it likely that these and other species will remain useful research models.

Because they are not defined as animals under the United States Department of Agriculture (USDA) Animal Welfare Act and Regulations, avian species are not included in publicly available USDA annual reports, and hence, the exact number of birds used in research annually in the United States is difficult to ascertain. In the United Kingdom, however, the Home Office reported that 4% of the 4.03 million (~160,000) animals used were birds, and this represented an increase of 14% over the previous year. By way of comparison, this is on par with the 147,112 hamsters reported in the USDA 2012 annual report.

This handbook was written to serve as a basic reference for those technicians, researchers, and veterinarians who have little experience working with birds in the research setting. The six chapters contained herein address general avian biology and physiology, husbandry, regulations that apply to the care and use of birds in research, experimental methods, and veterinary care. There are several comprehensive works regarding avian medicine and surgery, and those are cited throughout. Additionally, Chapter 6 provides a comprehensive list of resources that might be useful to those working with birds.

authors

Douglas K. Taylor has worked with a variety of avian species in varied settings since 1995. From 1995 to 2000, he worked in private practice, routinely managing avian medicine and surgery cases; during 2000–2002, he worked in the field of wildlife toxicology and used chickens to study PCB (polychlorinated biphenyls) and mercury toxicity; between 2002 and 2006, he was in training as a resident at the University of Michigan where pigeons, passerines, and chickens were in use. He is currently a faculty veterinarian at Emory University where a substantial number of passerine species and, occasionally, chickens are housed. He has been a diplomate in the American College of Laboratory Animal Medicine since 2006.

Vanessa K. Lee has worked with avian species in both a research and clinical environment since 1999. From 1999 to 2000, Dr. Lee worked as a veterinary technical assistant in the Exotics Department of the University of Georgia Small Animal Veterinary Teaching Hospital; during 2000–2001, she was an aviary farm manager for over 400 psittacine birds; between 2001 and 2004, Dr. Lee was a student research assistant for Dr. Branson Richie, working with pigeons and psittacines; from 2001 to 2003, she worked as a wildlife treatment crew volunteer and supervisor, with various wildlife species and poultry; and during 2005–2007, she was an associate veterinarian in private practice with a heavy avian caseload that included psittacines, poultry, passerines, ducks, pigeons, doves, and raptors. Dr. Lee did her two-year residency between 2007 and 2012 and subsequently held a faculty position at Emory University, during most of which she had clinical responsibilities for multiple passerine species.

She became a diplomate in the American College of Laboratory Animal Medicine in 2010.

Karen R. Strait has worked with avian species in a variety of settings since 2003. From 2003 to 2004, she was a wildlife treatment crew volunteer, working with various wildlife species and poultry. During 2004–2005, she completed a six-week training at the Southeastern Cooperative Wildlife Disease Study during which she participated in field studies and diagnostic necropsies of various avian species, including crows, passerines, and raptors. During 2005–2007 and 2011–2012, she did a two-year residency and subsequently held a faculty position at Emory University, during which she provided clinical support for multiple passerine species and poultry. She became a diplomate in the American College of Laboratory Animal Medicine in 2009.

important biological features

General Taxonomy. All birds are vertebrates belonging to the class Aves. There are an estimated 10,000 species of birds in approximately 45 orders. They are globally distributed and found in a variety of ecosystems from mountainous to coastal, forested, desert, and grasslands. The greatest diversity of avian families is found in South America and Australia. As a group, birds exhibit a remarkable array of phenotypic traits, varying widely in shape, size, behavior, and flight abilities. One needs only to compare and contrast a flightless penguin to a peregrine falcon to appreciate the diversity of this group.

It is generally accepted that birds descended directly from dinosaurs, although there is some debate about their precise placement on phylogenetic trees with some evidence suggesting that they are best considered a subgroup of reptiles instead of a class unto themselves. Regardless of those details, birds clearly share distinct characteristics with their reptilian ancestors such as the laying of amniote eggs, the presence of claws, and scales. In contrast to those similarities, birds evolved further to develop feathers, changes in bone structure and the respiratory system to enable flight, became endothermic, and replaced teeth with a hard beak. Subsequent to these early evolutionary steps, birds continued to develop traits that have enabled them to occupy a variety of niches today.

orders commonly used in research

Anseriformes

The order Anseriformes includes all species of ducks, geese, and swans. Although they are not widely used as research models, members of this order, primarily duck species, have been used with some regularity. The mallard duck (*Anas platyrhynchus*) has been used in studies of wildlife toxicology and was used to determine the etiology of avian vacuolar myelinopathy (Birrenkott et al., 2004), for example. Ducks have been used to study human hepatitis B virus pathogenesis as both the human and duck hepatitis B virus are substantially similar in structure and biology (Marion et al., 1984).

Galliformes

The order Galliformes is comprised of over 250 species of "chicken-like" birds and includes chickens, turkeys, quail, partridges, and all their relatives. Within this order, two species in the family Phasianidae are routinely used in research today, the domestic chicken (*Gallus gallus domesticus*) and the Japanese quail (*Cotornix japonica*). Chickens were the animal model in which the characterization of the B-lymphocyte (Cooper et al., 1966) and the discoveries of retroviruses and oncogenes (Svoboda, 1986) occurred. They are currently used primarily in studies of infectious disease and a variety of genetic disorders, hearing research, and some cancer studies. The Japanese quail and other closely related quail species are commonly involved in studies of environmental toxicants such as pesticides (Wilhelms et al., 2006). There has been historical interest in studying turkeys (*Meleagris* spp.) as a model of cardiovascular disease as they develop spontaneous and induced cardiomyopathy (Czarnecki et al., 1974), but they are not widely used.

Columbiformes

The order Columbiformes includes approximately 308 species of doves and pigeons. The common pigeon or rock dove (*Columba livia*) is the member of this order most commonly used in research today, typically in behavioral studies. Relatively recently, they have been considered as a model useful for the study of osteoarthritis (Rothschild and Panza, 2006). The white carneau variety was used historically in studies of atherosclerosis.

Passeriformes

The order Passeriformes is the largest containing over 5000 species. These are most commonly referred to as the "perching birds," a subgroup of which is comprised of the "songbird" species also referred to as the oscine clade. Songbird species are of great interest to those studying language and learning because they learn complex song patterns in order to communicate in a way similar to how humans learn to speak and communicate. Among the most commonly used species for such studies are the Bengalese finch (*Lonchura striata domestica*), zebra finch (*Taeniopygia guttata*), and European starling (*Sturnus vulgaris*) (Lee et al., 2013; Okanoya, 2004).

Psittaciformes

The order Psittaciformes consists of over 350 species of birds known more commonly as the parrots. There are no members of this order that could be considered a model of human conditions, although there has been some effort made to identify similarities in feather picking in psitaccine species and obsessive–compulsive behaviors in humans (Bordnick et al., 1994). They are mentioned here because studies of anesthesia and analgesia in the veterinary school setting using species such as the Hispaniolan Amazon parrots (*Amazona ventralis*) (Sanchez-Migallon Guzman et al., 2013) are not uncommon and the laboratory animal professional should have some familiarity with them as a group.

anatomy and function

An exhaustive discussion about avian anatomy and physiology is beyond the scope of this work. In addition, there can be substantial variation in some structures between species, which cannot be detailed here. This section will focus on and serve to highlight unique avian anatomy and make brief mention of the function of most structures. For more intensive discussions, the reader is directed to the reference list at the end of the chapter and to Chapter 6: Resources.

respiratory system

The respiratory system of birds is considered to be the most efficient of all vertebrates. Although form and function very closely mimic those of other species, there are anatomical features of the avian system that are entirely unique. The interested reader is referred to Brown et al. (1997) for a review of avian respiratory anatomy and function. Birds have paired nares that are typically located at the base of the beak.

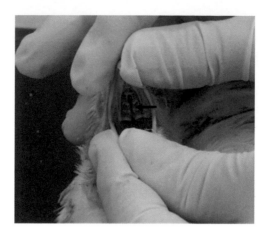

Fig. 1.1 Photograph showing the choanal slit (arrow) in the hard palate in the oral cavity of a chicken.

In some species such as budgerigars and pigeons, the nares are surrounded by a thick, soft tissue called the cere, which can be the site of disease and also can have pigmentation useful in sex determination. Inside the oral cavity lies a fissure in the hard palate called the choana or choanal slit (Figure 1.1). This opening creates a communication between the oral and nasal cavities and is thought to increase respiratory efficiency by minimizing airflow obstruction. The glottis (Figure 1.2) serves as the separation between the larynx and the trachea. This slit-like opening is readily visible at the base of the tongue in most species when the mouth is opened.

The trachea of birds contains complete cartilaginous rings that are ossified in passerine species and a few others. This is important to note when intubating, and details about intubation can be found in Chapter 4: Veterinary Care. The syrinx lies where the trachea meets the left and right primary bronchi. The structure of the syrinx is highly variable between species, but in most is composed of ossified cartilages, soft structures, membranes, and musculature. The syrinx is considered to be the vocal organ and is used to produce a remarkable array of sounds, particularly in passerine species (Trevisan and Mindlin, 2009).

Primary bronchi continue to branch into medioventral, mediodorsal, lateroventral, and laterodorsal secondary bronchi. Secondary bronchi branch further into parabronchi, and this network essentially forms the lung tissue. Lung tissue in birds is classified as either paleopulmonic or neopulmonic, with all species of birds

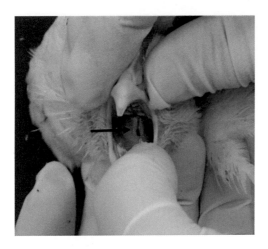

Fig. 1.2 Photograph showing the glottis (arrow) in the oral cavity of a chicken.

possessing the former, and only some possessing the latter. There are small cavities, the atria, which open to the parabronchi and lead to air capillaries. Air capillaries within the parabronchi are intertwined with blood capillaries, and gas exchange occurs at this level. The avian respiratory system has a volume several times larger than that of mammals for individuals of comparable size (Lasiewski and Calder, 1971).

Associated with the lung both anatomically and functionally is the system of air sacs. Most species possess eight air sacs, the singular cervical and clavicular, and the paired cranial, caudal, and abdominal. Air sacs do not participate in gas exchange but serve primarily to move air through the lung tissue. When healthy, the membranes of these sacs should be thin and transparent. Air sacculitis caused by fungal or bacterial agents occurs with some frequency in pet birds and might be a consideration when a bird presents clinically. Birds do not possess a diaphragm and breathe by expanding the body cavity and air sacs, which mandates avoidance of excessive body compression during handling of small species like many passerines. Interestingly, some air enters the caudal air sacs and some the cranial upon inspiration with some air then traveling from caudal to cranial air sacs at expiration. Because of the network of air sacs and parabronchi, two full breaths for air can be required to completely circulate through the respiratory tract (Lasiewski and Calder, 1971).

gastrointestinal system

The gastrointestinal system is not entirely different from that of mammalian species but does have some special anatomical features worthy of note. Overall, the digestive tract is shorter in birds compared to mammals. From the oral cavity, the esophagus enters the crop (Figure 1.3), which is essentially a dilatation of the esophagus at the thoracic inlet that serves purely for storage of food. When force feeding or in the case of parents feeding young, food is placed directly into the crop. The crop can be a site of disease such as fungal infections. It can also be injured when force feeding either mechanically or thermally if food items are warmed, so care must be taken to avoid these negative consequences. From the crop, the esophagus continues to the proventriculus, the glandular portion of the stomach akin to the typical mammalian stomach (Figure 1.4). The ventriculus, commonly referred to as the gizzard, has very thick, muscular walls and no glands (Figure 1.4). The lining consists of a tough membrane called the cuticle, which offers protection from secretions of the ventriculus and injury due to the grinding action of the organ. The primary function of the ventriculus is to mechanically break down food,

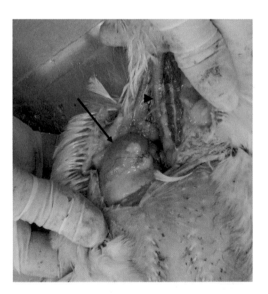

Fig. 1.3 Photograph showing the crop (arrow) of a chicken with skin reflected. Note that saline has been instilled to allow easier visualization. The trachea (arrowhead) is identified as an anatomical reference point.

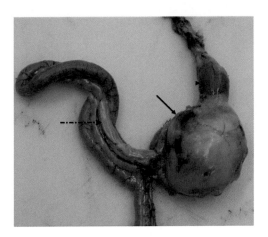

Fig. 1.4 Photograph showing the proventriculus or glandular stomach (arrowhead), ventriculus or gizzard (solid arrow), and pancreas (dashed arrow) lodged between duodenal loops from a chicken.

particularly hard items like some seeds, in preparation for complete digestion and absorption. In the wild, birds can be seen consuming small rocks, which serve as "grit" to help grind food in the gizzard. Grit can be provided to captive birds to perform the same function, although it is not necessary for some species. The remainder of the gastrointestinal system is generally similar to that of mammalian species, with the pylorus marking the separation of the ventriculus and duodenum, which then leads to the jejunum and ileum, which ultimately terminates in the cloaca, along with the ureters and genital ducts. The region of the cloaca, where the rectum enters the coprodeum, serves as the area where feces and liquid waste are stored until released. The opening of the cloaca, which is visible externally, is called the vent (Figure 1.5). The vent should always be viewed during physical exam and should be clean and without significant fecal material adhered to the feathers in the area.

The pancreas is typically found within the duodenal loops (Figure 1.4) and consists of three lobes. The exocrine pancreas secretes amylase, lipase, proteolytic enzymes, and sodium bicarbonate, which travel via one to three pancreatic ducts into the distal aspect of the ascending duodenum. The liver consists of two lobes that are subdivided further in some species. Right and left hepatic ducts combine to form the common hepatoenteric duct, which enters the duodenum. The gall bladder is connected to the right hepatic duct via the hepatocystic duct, and the bladder drains into the duodenum via the

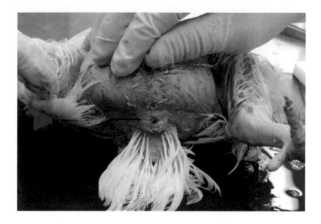

Fig. 1.5 Photograph of the vent (arrow) in a chicken.

cysicoenteric duct. In species lacking a gall bladder (e.g., pigeons), however, a branch of the right hepatoenteric ducts enters directly into the duodenum.

urogenital system

The general function of the urogenital system in birds is excretion and reproduction, which is not different from other species. Significant anatomical differences, however, do exist. The kidneys, as in most species, serve as a primary organ for nitrogenous waste filtration and excretion. In birds, the excreted product is uric acid, which is in contrast to urea produced by the mammalian kidney. Uric acid forms as a white sludge that is carried via the ureters from the kidney along with liquid urine to the cloaca. Birds lack a urinary bladder, and urine along with uric acid empties into the cloaca in the region known as the urodeum and is then stored in another region, the coprodeum, until released.

The ovaries of birds are found bilaterally in the abdomen just as would be the case for other species. The function and hormonal control of the ovarian follicle development is likewise similar to that of other species. As birds become sexually mature, however, it is only the left oviduct that develops fully with the right becoming vestigial and nonfunctional in many species. Consequently, all eggs develop from within the left oviduct, which leads to the uterus where shell glands secrete the hard, outer shell characteristic of all bird eggs. It should be noted that fertilization from a male is not necessary for egg development, and so some species of birds will lay eggs seasonally and provision of nest boxes can be a necessary part of husbandry and management. The testes of the male are found in the body cavity at the anterior pole of the kidneys. Both ovaries and testes in some species can undergo significant changes in pigmentation under hormonal influence during the course of a year, so it is important to define normal for the species at hand.

cutaneous/integumentary system

There are several notable anatomical features associated with the integumentary system of birds, some of which are highlighted here. Compared with mammals, the skin of birds is very thin. This attribute can be helpful when performing some procedures like venipuncure as vessels are fairly easy to visualize, but it also makes the skin susceptible to tearing when collecting blood and presents challenges to suturing. Birds have no sweat glands in their skin, so thermoregulation is accomplished primarily through the respiratory system and heat transfer from the surface. To increase or decrease heat transfer, birds will lift their wings away from their bodies and increase respiratory rate, or ruffle their feathers and crouch to the ground, respectively. These behavioral maneuvers are easily observed and can provide insight into the well-being of the animal. The only glands that most bird species do possess in their skin are the uropygial gland, also called the preen gland, holocrine glands near the external ear, and vent glands (Salibian and Montalti, 2009). The uropygial gland is devoid of feathers, and its primary function is to secrete oily, sebaceous material that is then spread across the feathers to create a waterproof barrier and keep the feathers in good condition. The secretions also contain normal bacterial flora that are thought to offer protection against pathogenic strains of bacteria (Ruiz-Rodriguez et al., 2009). The gland can be the site of some pathology such as tumors and infections. The functions of the ear and vent glands are not fully understood.

The beak of birds is composed of keratin sheaths. The shape of the beak is highly variable between different bird families. It is important to control the head and beak adequately when handling some birds like psitaccines as species in this order are capable of inflicting significant injury to the handler through biting. Animal care staff should also be aware that some birds will require routine beak trimming as part of their husbandry and management. In contrast to most of the body, the legs of birds are covered in thick scales and devoid of feathers. The claws of birds are also composed of keratin and can also require trimming as part of routine management should they become overgrown. Like the beak, the claws of larger species can cause significant injury if not properly restrained during handling.

The most noteworthy attribute of the integumentary system and defining of birds as an animal class is of course the presence of feathers. A complete discussion of feather structure is beyond the scope of this work. The focus here is on features important to those managing avian colonies. Feathers fall into three broad categories—contour, plume, and semiplume—based on structure. Contour feathers are the major feathers that are most visible externally, with plume feathers, also called down feathers, and semiplume feathers underlying the contours. Birds kept in captivity will completely lose and replace all feathers—a process called molting—typically once per year in association with breeding season, although some feathers can be replaced rather randomly, constituting an incomplete molt. The other exception to the general molting pattern is the loss of feathers on the abdomen or breast to form a bare area called the incubation or brood patch. This patch grows new feathers with the subsequent complete molt. The molting periodicity is somewhat dependent on species and age with molting occurring 2–3 times in young birds during the first year of life as they develop adult plumage. During molting, feathers are lost and replaced systematically so that the bird is never left bare or without the ability to fly and the entire process will take about 2 months. As an old feather is lost, it is replaced by a new feather in the same follicle and these replacement feathers are called pin or blood feathers. Pin feathers have ample blood supply and hemorrhage can be substantial if these young feathers are broken or damaged, particularly if they are the large quill feathers of the wings or tail. Removal of a damaged pin feather is the only treatment, and a new feather will replace it in due time. Feathers in healthy birds should always appear rather sleek and well kept as a result of preening behavior.

musculoskeletal system

The general structure and function of the avian musculoskeletal system is very similar to that of other vertebrate species but with some modifications that are largely reflective of their ability to fly. The bones of birds fall into two categories, the pneumatic and medullary. Pneumatic bones are filled with air and have connections with the air sac system. The skull, vertebrae, pelvis, sternum, ribs, humerus, and in some species, the femur, fall into this category. All other bones are medullary with large cavities and thin cortices. The medullary cavities have a network of spicules to create strength while minimizing weight. Compared to mammals, avian bones have a high mineral content resulting in high tensile strength but increased brittleness, which can present challenges to fracture fixation.

Avian skeletal muscle fibers can broadly and generally be classified as either white or red. White muscle, also known as "fast twitch" muscle is similar to mammalian muscle tissue in structure and innervated focally (Gauthier et al., 1982). Red muscle or "slow twitch," contains more mitochondria than white as well as more extensive capillary networks and myoglobin, which is responsible for the red color. Most muscle groups contain varying proportions of both muscle fiber types and the amount of each varies with developmental stages and in some cases, the reproductive cycle. Avian muscle contains the same actin and myosin filaments and contractile proteins troponin, tropomyosin, and alpha-actinin as mammals. The processes of excitation and contraction are therefore thought to occur in the same way as mammals.

The most apparent and unique feature of the avian appendicular skeleton is the wing. This appendage is homologous to the pectoral limb of mammals and indeed contains the same bones. The humerus, radius, and ulna are generally similar in appearance to those of mammalian species, while the carpal bones, metacarpal bones, and digits are more highly adapted to enable flight (Seki et al., 2012). The humerus is connected to the clavicular air sac. The ulna is larger than the radius and the distal aspect can serve as a location useful for intraosseous fluid administration. The bones of the pelvic limbs are likewise the same as those found in mammals. The digit arrangement varies widely among species and largely reflects the primary function of the feet, for example, perching, running, wading, paddling, or hunting.

The other noteworthy skeletal feature of birds is the shape of the sternum, which ventrally comes to a point in the shape of a boat's keel, hence, the sternum is also called the keel bone (Figure 1.6). The keel creates a large area for attachment of the substantial pectoral musculature, an adaptation almost solely to enable flight (Figure 1.7). Because this muscle mass is so substantial, it is a preferred site for administering intramuscular injections. The keel and associated

Fig. 1.6 Photograph showing the keel (arrow) in a chicken with the pectoral musculature reflected.

pectoral muscles should always be palpated as part of a thorough physical examination. The keel should be palpable, but not prominent in a well-conditioned bird. A keel that protrudes is typically the result of muscle atrophy and suggestive of a chronic disease process with associated loss of body condition.

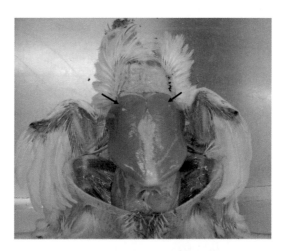

Fig. 1.7 Photograph showing the heavy pectoral musculature (arrows) in a chicken with the skin reflected.

cardiovascular system

The typical avian heart has four chambers with valve and chamber anatomy similar to that of mammals. The heart rate of birds is, on average, much higher than most other species with resting rates reaching over 700 beats per minute in some small passerines, although relative to body mass it is actually lower. Additionally, stroke volume, cardiac output, and heart mass are much higher compared to mammals per body mass unit, with hummingbirds having the largest (Grubb, 1983). All of these attributes exist presumably due to the high metabolic demands of flight. In fact, some migratory birds develop hypertrophy prior to migration seemingly in preparation for the rigorous exercise that the activity requires (Bishop and Butler, 1995). The cardiac electrophysiology of birds is similar to that of mammals, but there appears to be no agreement about what constitutes a normal avian electrocardiogram calling into question its value as a diagnostic or monitoring tool (Sturkie, 1986). Sympathetic neural control of heart rate is similar to many other species. Cardiac muscle responds to the catecholamines epinephrine and norepinephrine showing expected inotropic and chronotropic effects through beta-adrenergic receptors, although in birds the tissue is more responsive to the latter. Parasympathetic control likewise occurs as would be expected with acetylcholine serving as the neurotransmitter acting on muscarinic receptors with M1 receptors predominating, which differs from mammals where M2 is typically most abundant (Jeck et al., 1988).

Only the right jugular vein is prominent and available for blood collection while the left is vestigial or virtually absent. Other prominent vessels that are easily visualized and can be used for blood collection include the ulnar vein, which passes superficially over the ventral surface of the distal ulna and the median metatarsal vein, which runs along the medial aspect of the distal leg (see Chapter 5: Experimental Methodology). Birds are similar to reptiles in that they have both renal and hepatic portal systems (Akester, 1967; Sturkie and Abati, 1975). This feature becomes important when administering medications as there can be significant "first pass" effects for drugs heavily metabolized by the liver or kidney. Blood returning to the heart from approximately the caudal one-third of the bird circulates to the kidneys. Birds also have an efficient countercurrent system of blood flow in the tarsal and axillary region, which provides a mechanism for body heat conservation (Ederstrom and Brumleve, 1964).

hematology and immune system

The erythrocytes of birds are ovoid and nucleated much like those of reptiles. Red blood cells survive approximately 30 days in circulation, a relatively short time compared to 120 days in most mammals. Birds also have nucleated thrombocytes instead of platelets and neutrophils are replaced by heterophils as would be seen in rabbits. The cloacal bursa, also called the bursa of Fabricius, is a small outpouching on the proctodeum region of the cloaca that is unique to birds. The bursa along with the thymus are considered to be the major lymphoid organs in the bird with B-lymphocytes maturing in the former and T-lymphocytes maturing in the latter (Cooper et al., 1966). Both the bursa and the thymus spontaneously regress to some degree as birds age (Ciriaco et al., 2003). Avian B-lymphocytes produce the same immunoglobulins as mammals. One immunoglobulin has been called IgY by some because while it functions just as mammalian IgG, it has a different carbohydrate composition and also seemingly fills some of the role of IgE (Leslie and Benedict, 1970). There are homologues to both mammalian CD4 + helper and CD8 + killer T-cells in birds (Chan et al., 1988). Macrophages and natural killer cells are also present in birds, and their function is essentially the same as in other species.

In addition to the bursa of Fabricius and thymus, there are several other organs with significant lymphoid tissue. The spleen is an oval organ located adjacent to the liver. It contains white and red pulp like mammals. The cecal tonsil is a patch of tissue located at the proximal region of the cecum (Glick, 1979). It contains T-cells, B-cells, and plasma cells, functioning as a sort of sentinel organ given its position adjacent to the gastrointestinal tract. Peyer's patches exist along the intestine cranial to the ileocecal junction (Makala et al., 2002). Meckel's diverticulum or yolk stalk appears to contribute to the circulating pool of white blood cells (Olah and Glick, 1984). Lymph nodes are typically absent except in a few aquatic species.

endocrine system

The general organization and function of the avian endocrine system is similar to that of other species. The pituitary gland lies near the hypothalamus at the base of the brain near the optic chiasma. Tumors of this gland arise with some regularity, particularly in budgerigars with space-occupying adenomas being reported as most common followed by carcinomas (Langohr et al., 2012). It is comprised of two regions, the adenohypophysis and neurohypophysis. The adenohypophysis is separated into two parts, the pars distalis containing the secretory cells, and the pars tuberalis. Birds do not possess a pars intermedia like mammals. The adenophypophysis produces many of the same hormones found in other species including adrenal corticotrophic hormone (ACTH), melanin stimulating hormone (MSH), growth hormone, the gonadotropins—luteinizing and follicle stimulating hormone, thyroid stimulating hormone, and prolactin. Interestingly, prolactin stimulates the production of "crop milk" in both male and female birds. This "milk" is a mixture of sloughed crop wall mucosa, lipid, and protein that is fed to hatchlings for the first few days of life (Horseman and Buntin, 1995). Prolactin has also been shown to promote brooding and egg incubating behavior in birds (El Halawani et al., 1984).

The neurohypophysis of the pituitary gland consists of the pars nervosa, infundibular stalk, and the median eminence. The median eminence produces arginine vasotocin and mesotocin. Vasotocin is analogous in function to mammalian antidiuretic hormone and mesotocin is analogous to oxytocin (Goldstein, 2006; Goodson et al., 2009). As might be expected, receptors for these neuropeptides lie primarily in the kidney and uterus, respectively.

The thyroid glands are paired structures found inside the thoracic inlet near the carotid arteries. Their structure and function are similar to that of mammalian glands with thyroid hormones playing key roles in body weight regulation, plumage growth, fertility, lipid metabolism, and development of secondary sex characteristics (Merryman and Buckles, 1998). Thyroid hormone is secreted primarily as T4 (~60%) under the control of thyrotropin releasing hormone (TRH) produced in the hypothalamus which controls the release of thyrotropin stimulating hormone (TSH) produced in the pituitary gland. T3 is the biologically active form of the hormone. In adult birds, serum T3 levels appear to be the main controller of TSH production, not TRH (Merryman and Buckles, 1998). T4 binds weakly to albumin compared to mammals, hence, free T4 levels tend to be higher in avian species.

Birds have either two or four parathyroid glands just caudal to the thyroid glands. Parathyroid hormone (PTH) plays a key role in calcium homeostasis as it does in mammals. The major stimulus for PTH secretion is a fall in plasma calcium levels, which can be especially dramatic during egg shell formation. In response to these decreasing levels, PTH is secreted and calcium is mobilized from the skeleton to increase plasma levels. Birds appear to be much more sensitive to PTH, likely because of the demands of egg laying. Lying just caudal to the parathyroid glands are the ultimobranchial glands. These glands are rich with C-cells which secrete calcitonin. Unlike mammals, birds appear to be relatively unresponsive to calcitonin and this hormone seems to play little, if any, role in calcium regulation.

The adrenal glands are paired and located anterior and medial to the kidneys. Their general structure, function, and secretory control seems very similar to that of other species. Microscopically, they exhibit some zonation in cellular arrangement, but it is not as distinct as that seen in mammals (Aire, 1980; Holmes and Cronshaw, 1984). Corticosterone is the major corticosteroid produced with aldosterone produced in significantly lesser quantities. Birds possess a well-described hypothalamo–pituitary–adrenal (HPA) axis, and as part of that feedback system ACTH produced in the pituitary gland is the primary regulator of corticosterone release. Adrenocortical activation followed by corticosterone release as part of HPA axis function is a hallmark of the general stress response in birds. Other adrenocortical hormones include significant amounts of testosterone and estradiol (Tanabe et al., 1979).

The chromaffin cells of the adrenal glands secrete the catecholamines norepinephrine and epinephrine. These cells are the sole producers of epinephrine in birds while norepinephrine is also produced in postganglionic sympathetic nerve tissue in significant amounts. Catecholamine release is primarily under the influence of external factors such as acute stress, which activates the sympathochromaffin axis, but ACTH is also capable of stimulating secretion (Zachariasen and Newcomer, 1974).

The avian pancreas functions in much the same way it does in other species, serving as the organ largely responsible for glucose regulation. As mentioned earlier, it is situated between the ascending and descending loops for the duodenum. The endocrine pancreas is larger proportionally than the exocrine compared to mammals, but the islets consist of the same cell types with the A-cells secreting glucagon, the B-cells secreting insulin, D-cells somatostatin, and F-cells pancreatic polypeptide. Precisely how the endocrine pancreas controls glucose is not completely known. The primary trigger for insulin release, for example, is not glucose, and instead is driven largely by cholecystokinin, glucagon, and various amino acids. Interestingly, most birds are in a constant state of hyperglycemia by mammalian standards with commonly seen nonfasting levels of 250 mg/dL.

The pineal gland secretes melatonin, which is instrumental in control of circadian rhythm and photoperiodism in birds (Gwinner et al., 1997). The gland is located on the dorsal surface of the brain between the two hemispheres of the telencephalon and cerebellum. Most species possess a well-developed gland, but a few nocturnal species such as owls have a more vestigial gland. Regardless of whether a species is diurnal or nocturnal, melatonin secretion typically increases during the night and decreases during the day. The level of dependency upon the pineal gland as a mediator of circadian rhythm varies among species, but external factors, most importantly light intensity, seem to be universally important in driving circadian rhythm in avian species (Cassone and Menaker, 1984), highlighting the importance of environmental control within the animal facility.

nervous and sensory systems

The general anatomy and function of the avian nervous system and associated sensory systems are not appreciably different from those of most other species. A comprehensive discussion of these systems is beyond the scope of this book, and readers are referred to other works such as those by P.D. Sturkie for more detail. The hind and midbrain of birds are virtually identical in structure to mammalian species, while the forebrain of birds lacks a well-developed neocortex and cells homologous to those composing the neocortex are found instead in a region called the telencephalic complex. The midbrain has a substantial mesencephalic colliculus or optic lobe, which would be reasonably expected given the heavy reliance on vision. Birds possess 12 pairs of cranial nerves.

The spinal cord is covered by three meninges, and the number of spinal nerves varies with the number of vertebrae—from 38 in the pigeon to 51 pairs in the ostrich. The spinal cord and neural tube are of the same length so nerves pass laterally through their foramen and there is no cauda equinae. The white matter of the spinal cord consists of the dorsal, lateral, and ventral columns, and the gray matter is composed of dorsal and ventral horns. Ascending and descending pathways have been characterized and are seemingly very similar to those of mammals, importantly suggesting similar somatosensory capabilities for pain.

Birds possess all the senses of other species and, except for features noted here, structure and function of sensory organs are not particularly unique. Avian species tend to rely heavily on visual acuity to function, and indeed, the eyes are larger than other species. The combined weight of the eyes is often more than the brain in an individual, and ostriches possess the largest eye of any terrestrial vertebrate (King and McLelland, 1984). The ciliary muscles are striated in contrast to the smooth muscle found in mammals, and the mechanism of accommodation is different. In mammals, ciliary muscles relax to allow the lenses to assume a more spherical shape while in bird, the muscles actually push against the lenses to achieve the shape change in coordination with iris sphincter muscles thus allowing excellent accommodation capabilities even to the point of bulging through the pupil (Levy and Sivak, 1980; Sivak, 1980). The pupils of most birds are round and like the ciliary muscles, the sphincter and dilator muscles are striated. The lens of bird eyes is much softer than that of mammals and allows transmission of a wider spectrum of light wavelengths (King and McLelland, 1984). The retina is relatively thick and lacks blood vessels. The proportion of rods and cones varies with the lifestyle of the species; for example, nocturnal birds such as owls have almost all rods and few cones. There are several features of the avian retina thought to increase visual acuity significantly. The retinal central fovea of birds is deep, some species also have a lateral, temporal fovea (Ruggeri et al., 2010), and there is approximately two times the concentration of cones per square millimeter. The result is that a pigeon, for example, can distinguish 140 flashes per second compared to 70 for a human. It is presumed, based on the fact that brightly colored plumage is used to communicate and influences behavior, that birds possess color vision. Protruding from the retina into the vitreous is a pigmented structure called the pecten. Because it is highly vascularized, it is generally accepted that the primary role of the pecten is to provide nutrients to the avascular retina, while other functions, such as serving as a protective organ, are the subject of some debate (Barlow and Ostwald, 1972). The nictitating membrane is present, and its movement is controlled by two striated muscles, the quadratis and pyramidalis. In most species, the membrane is transparent, and it is thought that birds can fly with the membrane over the eye to protect against dessication. It likely serves a similar protective role in diving birds as well.

Birds lack ear pinnae, and specialized contour feathers, the ear coverts, help direct sound into the meatus. In owls, the feathers of the facial disc also help direct sound to the meatus. The tympanic membrane of birds projects outward instead of inward. The columella, homologous to the mammalian stapes, serves to transmit vibration from the tympanic membrane to the vestibular window (Arechvo et al., 2013). The cochlea and cochlear duct are relatively short, but the receptor cell density is about 10 times that of mammals, so the total number of cells is likely similar. In terms of auditory capacity, it is thought that birds hear in a narrower spectrum of wavelengths, but that the resolution is superior to mammals.

The sense of smell in birds is thought to be relatively weak, the logic being that while flying at altitude at some speed, the stimuli would be very dilute and unlikely to serve a significant functional role. There are, however, studies in pigeons to suggest that olfaction is fairly well developed in birds, and certainly in some carrion-eating species such as vultures it is used heavily to locate food sources. In fact, a recent study of olfactory receptors suggests that birds possess a sense of smell better developed than originally thought (Steiger et al., 2008). Like the ability to discern odors, the avian sense of taste is believed to be poorly developed, and the role of taste in normal behavior is not fully understood. The number of taste buds varies by species and, interestingly in some cases, are not located on the tongue but instead are found in various areas of the oropharynx. In all instances, birds have significantly fewer taste buds than mammals, and therefore acuity is thought to be diminished. Studies of taste preference show that they vary widely with some species preferring sour flavors and others sweet (Duncan, 1963). Almost uniformly bitter and salty solutions are not rejected.

references

Aire, T.A. 1980. Morphometric study of the avian adrenal gland. *Journal of Anatomy.* 131:19–23.

Akester, A.R. 1967. Renal portal shunts in the kidney of the domestic fowl. *Journal of Anatomy.* 101:569–594.

Arechvo, I., Zahnert, T., Bornitz, M., Neudert, M., Lasurashvili, N., Simkunaite-Rizgeliene, R., and Beleites, T. 2013. The ostrich middle ear for developing an ideal ossicular replacement prosthesis. *European Archives of Oto-Rhino-Laryngology.* 270:37–44.

Barlow, H.B., and Ostwald, T.J. 1972. Pecten of the pigeon's eye as an inter-ocular eye shade. *Nature: New Biology.* 236:88–90.

Birrenkott, A.H., Wilde, S.B., Hains, J.J., Fischer, J.R., Murphy, T.M., Hope, C.P., Parnell, P.G., and Bowerman, W.W. 2004. Establishing a food-chain link between aquatic plant material and avian vacuolar myelinopathy in mallards (*Anas platyrhynchos*). *Journal of Wildlife Diseases.* 40:485–492.

Bishop, C., and Butler, P. 1995. Physiological modelling of oxygen consumption in birds during flight. *The Journal of Experimental Biology.* 198:2153–2163.

Bordnick, P.S., Thyer, B.A., and Ritchie, B.W. 1994. Feather picking disorder and trichotillomania: An avian model of human psychopathology. *Journal of Behavior Therapy and Experimental Psychiatry.* 25:189–196.

Brown, R.E., Brain, J.D., and Wang, N. 1997. The avian respiratory system: A unique model for studies of respiratory toxicosis and for monitoring air quality. *Environmental Health Perspectives.* 105:188–200.

Cassone, V.M., and Menaker, M. 1984. Is the avian circadian system a neuroendocrine loop? *The Journal of Experimental Zoology.* 232:539–549.

Chan, M.M., Chen, C.L., Ager, L.L., and Cooper, M.D. 1988. Identification of the avian homologues of mammalian CD4 and CD8 antigens. *Journal of Immunology.* 140:2133–2138.

Ciriaco, E., Pinera, P.P., Diaz-Esnal, B., and Laura, R. 2003. Age-related changes in the avian primary lymphoid organs (thymus and bursa of Fabricius). *Microscopy Research and Technique.* 62:482–487.

Cooper, M.D., Raymond, D.A., Peterson, R.D., South, M.A., and Good, R.A. 1966. The functions of the thymus system and the bursa system in the chicken. *The Journal of Experimental Medicine.* 123:75–102.

Czarnecki, C.M., Jankus, E.F., and Hultgren, B.D. 1974. Effects of furazolidone on the development of cardiomyopathies in turkey poults. *Avian Diseases.* 18:125–133.

Duncan, C.J. 1963. Response of feral pigeon when offered active ingredients of commercial repellents in solution. *Annals of Applied Biology.* 51:127.

Ederstrom, H.E., and Brumleve, S.J. 1964. Temperature gradients in the legs of cold-acclimatized pheasants. *The American Journal of Physiology.* 207:457–459.

El Halawani, M.E., Burke, W.H., Millam, J.R., Fehrer, S.C., and Hargis, B.M. 1984. Regulation of prolactin and its role in gallinaceous bird reproduction. *The Journal of Experimental Zoology.* 232:521–529.

Gauthier, G.F., Lowey, S., Benfield, P.A., and Hobbs, A.W. 1982. Distribution and properties of myosin isozymes in developing avian and mammalian skeletal muscle fibers. *The Journal of Cell Biology.* 92:471–484.

Glick, B. 1979. The avian immune system. *Avian Diseases.* 23:282–289.

Goldstein, D.L. 2006. Regulation of the avian kidney by arginine vasotocin. *General and Comparative Endocrinology.* 147:78–84.

Goodson, J.L., Schrock, S.E., Klatt, J.D., Kabelik, D., and Kingsbury, M.A. 2009. Mesotocin and nonapeptide receptors promote estrildid flocking behavior. *Science.* 325:862–866.

Grubb, B.R. 1983. Allometric relations of cardiovascular function in birds. *The American Journal of Physiology.* 245:H567–H572.

Gwinner, E., Hau, M., and Heigl, S. 1997. Melatonin: Generation and modulation of avian circadian rhythms. *Brain Research Bulletin.* 44:439–444.

Holmes, W.N., and Cronshaw, J. 1984. Adrenal-gland—Some evidence for the structural and functional zonation of the steroidogenic tissues. *Journal of Experimental Zoology.* 232:627–631.

Horseman, N.D., and Buntin, J.D. 1995. Regulation of pigeon crop-milk secretion and parental behaviors by prolactin. *Annual Review of Nutrition.* 15:213–238.

Jeck, D., Lindmar, R., Löffelholz, K., and Wanke, M. 1988. Subtypes of muscarinic receptor on cholinergic nerves and atrial cells of chicken and guinea-pig hearts. *British Journal of Pharmacology.* 93:357–366.

King, A.S., and McLelland, J. 1984. *Birds: Their Structure and Function*, second ed. (Philadelphia, PA: Bailliere Tindall).

Langohr, I.M., Garner, M.M., and Kiupel, M. 2012. Somatotroph pituitary tumors in budgerigars (*Melopsittacus undulatus*). *Veterinary Pathology.* 49:503–507.

Lasiewski, R.C., and Calder, W.A., Jr. 1971. A preliminary allometric analysis of respiratory variables in resting birds. *Respiration Physiology.* 11:152–166.

Lee, J.V., Maclin, E.L., Low, K.A., Gratton, G., Fabiani, M., and Clayton, D.F. 2013. Noninvasive diffusive optical imaging of the auditory response to birdsong in the zebra finch. *Journal of Comparative Physiology A: Neuroethology, Sensory, Neural, and Behavioral Physiology.* 199:227–238.

Leslie, G.A., and Benedict, A.A. 1970. Structural and antigenic relationships between avian immunoglobulins. 3. Antigenic relationships of the immunoglobulins of the chicken, pheasant and Japanese quail. *Journal of Immunology.* 105:1215–1222.

Levy, B., and Sivak, J.G. 1980. Mechanisms of accommodation in the bird eye. *Journal of Comparative Physiology.* 137:267–272.

Makala, L.H., Suzuki, N., and Nagasawa, H. 2002. Peyer's patches: Organized lymphoid structures for the induction of mucosal immune responses in the intestine. *Pathobiology.* 70:55–68.

Marion, P.L., Knight, S.S., Ho, B.K., Guo, Y.Y., Robinson, W.S., and Popper, H. 1984. Liver disease associated with duck hepatitis B virus infection of domestic ducks. *Proceedings of the National Academy of Sciences of the United States of America.* 81:898–902.

Merryman, J.I., and Buckles, E.L. 1998. The avian thyroid gland. Part one: A review of the anatomy and physiology. *Journal of Avian Medicine and Surgery.* 12:234–237.

Okanoya, K. 2004. The Bengalese finch: A window on the behavioral neurobiology of birdsong syntax. *Annals of the New York Academy of Sciences.* 1016:724–735.

Olah, I., and Glick, B. 1984. Meckel's diverticulum. I. Extramedullary myelopoiesis in the yolk sac of hatched chickens (*Gallus domesticus*). *The Anatomical Record.* 208:243–252.

Rothschild, B.M., and Panza, R. 2006. Osteoarthritis is for the birds. *Clinical Rheumatology.* 25:645–647.

Ruggeri, M., Major, J.C., Jr., McKeown, C., Knighton, R.W., Puliafito, C.A., and Jiao, S. 2010. Retinal structure of birds of prey revealed by ultra-high resolution spectral-domain optical coherence tomography. *Investigative Ophthalmology & Visual Science.* 51:5789–5795.

Ruiz-Rodriguez, M., Valdivia, E., Soler, J.J., Martin-Vivaldi, M., Martin-Platero, A.M., and Martinez-Bueno, M. 2009. Symbiotic bacteria living in the hoopoe's uropygial gland prevent feather degradation. *The Journal of Experimental Biology.* 212:3621–3626.

Salibian, A., and Montalti, D. 2009. Physiological and biochemical aspects of the avian uropygial gland. *Brazilian Journal of Biology = Revista Brasleira de Biologia.* 69:437–446.

Sanchez-Migallon Guzman, D., Braun, J.M., Steagall, P.V., Keuler, N.S., Heath, T.D., Krugner-Higby, L.A., Brown, C.S., and Paul-Murphy, J.R. 2013. Antinociceptive effects of long-acting nalbuphine decanoate after intramuscular administration to Hispaniolan Amazon parrots (*Amazona ventralis*). *American Journal of Veterinary Research*. 74:196–200.

Seki, R., Kamiyama, N., Tadokoro, A., Nomura, N., Tsuihiji, T., Manabe, M., and Tamura, K. 2012. Evolutionary and developmental aspects of avian-specific traits in limb skeletal pattern. *Zoological Science*. 29:631–644.

Sivak, J.G. 1980. Accommodation in vertebrates: A contemporary survey. *Current Topics in Eye Research*. 3:281–330.

Steiger, S.S., Fidler, A.E., Valcu, M., and Kempenaers, B. 2008. Avian olfactory receptor gene repertoires: Evidence for a well-developed sense of smell in birds? *Proceedings Biological Sciences/The Royal Society*. 275:2309–2317.

Sturkie, P.D. (ed.) 1986. Heart: Contraction, conduction, and electrocardiography. In *Avian Physiology* (New York, NY: Springer), pp. 167–190.

Sturkie, P.D., and Abati, A. 1975. Blood flow in mesenteric, hepatic portal and renal portal veins of chickens. *Pflügers Archiv: European Journal of Physiology*. 359:127–135.

Svoboda, J. 1986. Rous sarcoma virus. *Intervirology*. 26:1–60.

Tanabe, Y., Nakamura, T., Fujioka, K., and Doi, O. 1979. Production and secretion of sex steroid hormones by the testes, the ovary, and the adrenal glands of embryonic and young chickens (*Gallus domesticus*). *General and Comparative Endocrinology*. 39:26–33.

Trevisan, M.A., and Mindlin, G.B. 2009. New perspectives on the physics of birdsong. *Philosophical Transactions Series A, Mathematical, Physical, and Engineering Sciences*. 367:3239–3254.

Wilhelms, K.W., Cutler, S.A., Proudman, J.A., Anderson, L.L., and Scanes, C.G. 2006. Effects of atrazine on sexual maturation in female Japanese quail induced by photostimulation or exogenous gonadotropin. *Environmental Toxicology and Chemistry/SETAC*. 25:233–240.

Zachariasen, R.D., and Newcomer, W.S. 1974. Phenylethanolamine-N-methyl transferase activity in the avian adrenal following immobilization or adrenocorticotropin. *General and Comparative Endocrinology*. 23:193–198.

2

husbandry

introduction

Composed of more than 10,000 species of birds, the class Aves is extremely diverse. There is a dearth of information on the basic physiological and behavioral needs of many species, and the husbandry requirements of most are not well defined. The wide variation in species makes it impractical to offer specific recommendations in this work that will address all situations. Instead, the information provided herein contains general considerations and recommendations for animal facilities and is not meant to serve as a comprehensive guide for avian husbandry programs. When working with an unfamiliar avian species, the reader is advised to research the birds' natural environment and to mimic these conditions as closely as possible. See Chapter 6: Resources, for commercially available housing options and products.

The many components of a husbandry program including housing, identification, environmental monitoring, enrichment, nutrition, and sanitation are designed to meet the animals' physical as well as psychological needs. A broad overview of these topics and recommendations are provided in *The Guide for the Care and Use of Laboratory Animals* (aka The Guide) for biomedical research facilities and the *Guide for the Care and Use of Agricultural Animals in Research and Teaching* (aka The Ag Guide) (FASS, 2010; NRC, 2011) for agricultural research facilities; facilities should design their husbandry program with these concepts in mind. Additionally, many species are regulated by the Convention on International Trade in Endangered Species of Wild Fauna and Flora (CITES), and facilities should ensure compliance with this and other local and national regulations before beginning research. More information about the regulatory agencies governing avian research can be found in Chapter 3: Management.

housing

General Considerations for Avian Facilities

- Most well-designed animal facilities can easily house avian species with minor modifications to accommodate special research needs.

- As a primary consideration, all housing systems should be designed with animal comfort in mind so that birds can make normal postural adjustments, stay clean, dry, and free of disease, and live under species-appropriate environmental conditions.

- All housing types present trade-offs. For example, birds housed on litter may have greater opportunity to forage but run an increased risk of developing pododermatitis from standing on wet substrate (Lay et al., 2011). Birds housed in groups have more opportunity to interact with conspecifics but may have increased incidence of feather pecking (FASS, 2010). The choice of housing must be balanced between the positive and negative attributes of each system with consideration for the needs of the animals, facility personnel, and researchers.

- Most birds are capable of flight and many species fly long distances over the course of the day. Cages that provide the opportunity for flight are recommended when consistent with experimental goals.

- Similarly, cages should allow birds the opportunity to engage in species-specific behaviors such as perching, nesting, wing-flapping, and foraging.

- Different types of housing may be required for different life stages. For example, chicks require ambient temperature to be higher than that required for adults.

- Although it may be advantageous to house birds in rooms with windows to provide natural light, cages should be positioned away from windows and doors to prevent startling. Additionally, light impacts breeding and birds will adjust breeding behavior to the natural light cycle, rendering artificial breeding suppression or stimulation ineffective.

- In general, birds should be separated by species to avoid interspecies transmission of disease as well as anxiety caused by interspecies conflict.
- Birds of prey should not be housed near prey species.

Housing Types

Cages

Commercially manufactured cages for housing birds in the laboratory can be found that will accommodate the most commonly used species such as poultry, quail, pigeons, and finches (Figure 2.1). Cages for individual birds should be large enough to allow birds to stretch wings, turn comfortably, and permit natural behavior to the degree possible. Often cages designed for other species can be modified to house avian species. Individual and glove box isolators for infectious disease studies or where birds must be of a defined microbial status

Fig. 2.1 Rack of individual chicken cages.

are also available. Caging for agricultural studies can be purchased from traditional agricultural suppliers.

The materials used for cage construction must be nontoxic and easy to clean. Cages constructed with uncoated type 304 stainless steel are more resistant to corrosion from chemicals used in sanitation. Galvanized wire, commonly used in less expensive cages, can be the source of zinc toxicity in psittacines if materials are ingested (Howard, 1992). Cages made from this material should be initially washed with diluted acetic acid (vinegar) and brushed to remove clumps of oxidized zinc, also known as "white rust" (Howard, 1992).

Walls may be solid or mesh. Mesh walls allow birds to see conspecifics but cages must have enough space between them to prevent injury. Cages with three-sided solid walls and an open front are generally less stressful for nonsocial birds. Climbing species such as psittacines require mesh walls on which to grip and climb.

Floors may be solid or a grid design. Solid floors allow birds to be housed on substrate which encourages natural foraging activity, and there is evidence that domestic fowl prefer solid flooring (Dawkins, 1983; Hawkins et al., 2003). Cockatiels engage in "running" behavior and housing on solid floors is recommended (Kalmar et al., 2010). Grid floors, however, provide greater separation between the bird and its waste and allow easier sanitation since waste will fall through the wire into a pan. Grid floors may be constructed from polyvinyl chloride coated mesh wire or perforated to simulate wire while providing greater comfort. Provision of a perch or solid resting area and foraging board will encourage species-specific behavior and help reduce the incidence of foot problems. Cages designed specifically for laying hens have sloped floors to allow eggs to roll to the front for collection.

Waste collection systems may have removable pans or be designed to allow direct flushing with water. Waste in flush-based systems should be cleaned daily and care must be taken to avoid wetting birds. Waste pans made in a drawer design allow pans or liners to be cleaned without opening the main enclosure (Figure 2.2). Hanging style cages allow birds to be housed at a more comfortable height while facilitating sanitation under the cage. Cages with external access to feeders and waters facilitate husbandry procedures for personnel while minimizing disturbance to the birds.

Groups of poultry can be housed in a commercially available battery or furnished style cages. There is still considerable controversy over the use of conventional versus furnished poultry cages, and the European Union banned the use of conventional battery cages for commercial use in 2012. In contrast to furnished cages, conventional

Fig. 2.2 Drawer design waste pan on a songbird cage.

cages for hens lack nests, dust baths, and perches. While furnished cages allow hens to engage in more species-specific behavior, conventional cages are associated with decreased incidence of foot problems and keel deformities. A higher incidence of comfort behaviors (e.g., preening, wing flapping, etc.) has been observed in chickens housed in furnished cages suggesting this may be a favorable alternative (Pohle and Cheng, 2009). Although battery cages are permissible in the United States provided they meet the specifications outlined in the Ag Guide, modifications can be made in a laboratory setting to enhance animal welfare.

Pen

Ground-dwelling species such as poultry, Japanese quail, and pigeons can be housed in groups on the floor with substrate such as sand, straw, or wood shavings. The bedding should create as little dust as possible and must be fastidiously maintained to minimize wet litter and ammonia buildup. Pens allow for larger social groups of animals but can be more difficult to keep clean than cages. Typically, pens are not commercially available and must be constructed on-site. Large animal cages (e.g., canine or swine runs) can also be modified to house groups of birds (Figure 2.3).

Flight cage

Flight cages may be used for permanent group housing, or as exercise areas for birds normally housed in cages that do not permit flight. Cages that have a longer design typically allow for better flight. The addition of accessories, such as perches, foliage, and nest boxes, will

Fig. 2.3 Canine runs were modified to house these layer hens. In addition to bedding and perches (shown), a top panel was installed to prevent flight between runs. (Photo courtesy of Dorcas O'Rourke. With permission.)

provide a more complex, stimulating environment (Figure 2.4). Flight cages should be designed with species considerations in mind. For example, tree-dwelling species typically prefer to perch above the shoulder height of humans. Many ground-dwelling species have short, rapid takeoffs and may need protection from head trauma if startled. Disadvantages of this type of caging include difficulty in observing individual birds and increased difficulty in capturing birds.

Outdoor housing

Outdoor housing provides natural lighting, greater space for longer flights, and a more stimulating environment for species that can adapt to the local climate as benefits. Outdoor facilities must provide shelter from the elements as well as visual barriers or enclosures for birds to escape and hide. Combination indoor/outdoor housing gives birds a choice between more natural and lab-controlled conditions and can be partitioned off when outdoor conditions are not suitable. Rooftop housing is used by some facilities to provide outdoor access while limiting personnel access and reducing the threat of predators and some vermin. Since outdoor enclosures come with increased escape risks, a secondary enclosure, such as a large net, is necessary to contain escapees. In addition to the drawbacks described for flight cages, a wide range of environmental variables, increased potential for disease transmission from vermin, and increased risk of and liability associated with escape are inherent disadvantages associated with outdoor housing.

Fig. 2.4 Flight cage housing a colony of Bengalese finches. Perches, visual barriers, nest boxes, water baths, toys, and greens increase the complexity of the environment and promote species-specific behavior. The cage is attached to a tract on the ceiling to facilitate sanitation.

Brooders

Chicks are unable to regulate their body temperature for the first few weeks of life and require housing that provides additional thermal support. Poultry chicks can be housed in a floor pen or a box under a suspended electrical heater or alternatively in an electrical brooder (Figure 2.5). The temperature in the enclosure must be carefully

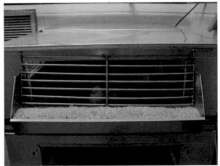

Fig. 2.5 Electrical poultry brooder.

monitored and adjusted as chicks mature. Electrical brooders typically have built-in radiant heaters on the roof of the cage that can be
controlled by a thermostat. The temperature of suspended heaters,
such as an infrared lamp, can be adjusted by altering the height of
the lamp. Chick behavior can be used to determine whether or not
the heat source is properly adjusted. Hot chicks are quiet and stand
away from the heat source, whereas cold chicks will huddle under
the heat to stay warm. Chicks that are kept at a comfortable temperature will be evenly dispersed and moderately noisy. Passerine
and psittacine chicks are altricial at birth and are under complete
parental care in nests posthatching.

Space recommendations

Few studies of the optimal housing space for avian species have
been conducted. In practice, adequate cage size will be defined by
a variety of factors such as species, group versus single housing,
number of birds, and breeding requirements. At a minimum, birds
must be able to comfortably engage in normal postural adjustments
and species-specific behavior such as wing flapping. Space recommendations for poultry and pigeons in the laboratory and poultry in
agricultural research are provided in the Guide and the Ag Guide,
respectively (FASS, 2010; NRC, 2011) (Table 2.1). Allowances for zebra
finches, ducks, and geese are given in Appendix A of The Council
of European Convention for the Protection of Vertebrate Animals
Used for Experimental and Other Scientific Purposes (CEC, 2006).

Table 2.1: Minimum Space Requirements for Single or Group-Housed
Avian Species

Species	Weight (kg)	Floor Area/Animal[a] ft²	m²	Height
Pigeons	–	0.8	0.07	Animals should be
Quail	–	0.25	0.023	able to comfortably
Chickens	<0.25	0.25	0.023	stand erect with their
	≤5.0	0.50	0.046	feet on the floor.
	≤1.5	1.00	0.093	
	≤3.0	2.00	0.186	
	>3.0	≥3.0	≥0.279	

Source: Adapted from NRC. 2011. *Guide for the Care and Use of Laboratory Animals,*
        8th edition. The National Academies Press, Washington, DC.
[a] Singly housed birds may require more space per animal than pair- or group-housed
   birds.

Recommendations for additional species including psittacines are found in Hawkins et al. (2001).

Provision of food and water

Food can be supplied in feeders or scattered on the floor to encourage foraging. A variety of feeder styles exist, including troughs, cups, bells, hanging feeders, and those designed to minimize food wastage. Petri dishes work well for species with precocial chicks (i.e., born with feathers and eyes open) that can eat on their own after birth.

Clean potable water should be available at all times. Water can be supplied in cups, bells, bottles, troughs, nipples, or drip cups (i.e., cups fitted under the nipple to prevent spillage). Many facilities use automatic waterers fitted with dishes, nipples, or the combination. Nipple systems are generally more sanitary than dishes, but birds must learn to drink from them. In general, poultry and psittacines will readily adjust to nipples. Songbirds, however, fare better with cage mounted gravity fed bottles or dishes, and pigeons, because they drink water by sucking through their beak, should be provided water in a dish. Pebbles can be placed in the water dishes of small chicks to prevent drowning. Waterers should be checked daily and cleaned as necessary to prevent fouling and clogging.

Food and water should be readily visible, especially to small chicks and birds placed in a new environment. Poultry chicks may need to be taught where to find and how to use a food or water source. If eating or drinking is not observed, tapping the dish or gently dipping the beak in the food/water will facilitate learning.

Feeders and watering devices can be attached to the sides of a cage or pen, hung from the ceiling, or placed directly on the floor. Ideally, these items should not be placed directly under perches to prevent bird droppings from soiling the containers. For flighted species, feeders and waterers can be attached to the cage at the end of the perch to allow birds a comfortable landing spot. Sufficient space must be provided to prevent competition from dominant birds. Minimum recommended feeder and waterer space for birds in agricultural research is provided in the Ag Guide (FASS, 2010).

Nutrition

Good nutrition is an essential part of avian health; however, providing a balanced diet in captivity can be challenging. In the wild, the avian diet varies greatly by species, geographic location, season, and life stage. Whereas some birds are omnivorous and consume a variety of seeds, plants, and insects, others are adapted as granivores, herbivores, or even carnivores. Complicating matters, some birds change from specializing in one type of food to another due to seasonal availability or breeding activity (Bateson and Feenders, 2010; Kalmar et al., 2010). With few exceptions (e.g., poultry), the specific nutritional requirements of most avian species are poorly defined at this time. Since inadequate nutrition can lead to a host of disease conditions in birds, the diet should be carefully researched prior to housing any new avian species and formulated to mimic as closely as possible the natural food preferences.

General considerations for avian nutrition include

- A good quality commercially formulated diet can be purchased from laboratory animal or zoo diet manufacturers and used as the basis of a nutritionally complete diet. Pelleted or extruded diets are available for several species and are generally recommended over seed mixtures in the laboratory setting. Pelleted/extruded diets provide consistent nutrition in each bite and prevent selective feeding (Schulte and Rupley, 2004). Seed diets, in contrast, are difficult to make nutritionally complete and some birds, such as parrots, will selectively consume some seeds (e.g., sunflower seeds) over others leading to nutritional imbalances (Kalmar et al., 2010).

- Dietary enrichment in the form of vegetables, fruits, nuts, seeds, or insects can be used to supplement the base diet. This enrichment serves two purposes: (1) to provide additional nutrients that are deficient in the base diet catered to the individual species and (2) to provide psychological stimulation by offering variety and encouraging foraging activity.

- When providing any new food, it is important to introduce the new item gradually and in conjunction with the current diet. Avian dietary preferences are shaped by early experiences; therefore, some birds will be wary of accepting a novel food item (Hawkins et al., 2003).

- Avocado and chocolate are toxic to birds and should be avoided.

- Energy requirements will vary by species, environment, life stage, and activity level. Many species of birds will adjust their food consumption to meet metabolic demands and can be fed ad libitum (Koutsos et al., 2001). Laying strains of chickens, for example, do well with ad libitum access to feed. Passerines fed ad libitum may consume less energy and maintain a lower body weight than if fed intermittently, and therefore low body fat should not necessarily be an indicator of poor welfare (Bateson and Feenders, 2010). In contrast, some species, most notably pigeons and psittacines, trend toward obesity when fed ad libitum, due to lack of exercise or a preference for high-energy foods (Fair et al., 2010; Kalmar et al., 2010). These birds often fare better on a restricted diet supplemented with produce and other lower calorie enrichment items to prevent boredom (Kalmar et al., 2010).

- Birds have varying nutritional needs through different life stages. Growing birds usually require higher energy and protein. Nutritionally complete commercial starter (chick) food is available in small crumbles or mash for precocial species. Altricial neonates require hand feeding with formula if they must be separated from the parent. Breeder diets with increased energy and calcium are available for breeding birds.

- Calcium is an essential mineral in the avian diet, and both calcium deficiency and excess can lead to skeletal abnormalities or reproductive problems. The dietary calcium content for growing precocial birds should be approximately 1% of dry matter, 0.5%–1% for adults, and up to 3.6% for laying hens (Hawkins et al., 2001). The provision of cuttlebones or oyster shells to species that seek out calcium should be considered. Calcium must be balanced with phosphorus to ensure adequate calcium availability.

- Vitamin supplements should be used with caution since some become toxic when consumed in excess. Seed mixtures are often sprayed with vitamins; however, birds that hull seeds (e.g., passerines and psittacines) will discard the coated exterior before consuming the seeds (Schulte and Rupley, 2004). Water supplementation is likewise unreliable since vitamins quickly break down in water, reduce water palatability, and may alter water consumption making accurate dosing difficult (Schulte and Rupley, 2004; Kalmar et al., 2010).

- The most common vitamin deficiencies among birds in captivity are Vitamins A and D. Deficiencies can be prevented through the provision of a balanced diet.

Grit

Grit is available in two forms: soluble and insoluble. Soluble grit, such a cuttlebone and crushed oyster shells, dissolves as it passes through the digestive tract and serves as a source of calcium in the avian diet (Taylor, 1996). Insoluble grit, generally found as sand or small pebbles, remains in the gizzard after ingestion and helps with the mechanical breakdown of food (Taylor, 1996). Some species of birds, particularly herbivorous or granivorous species such as galliformes, require insoluble grit in their diet for efficient digestion (Hawkins et al., 2001). Insoluble avian grit can be purchased from commercial suppliers and provided to birds as either a "pinch" or spread over the floor to prevent overconsumption and possible impaction (Fair et al., 2010). Birds will naturally choose the appropriate size of grit if a variety of sizes are provided (Hawkins et al., 2001). Insoluble grit is not necessary for psittacines and passerines because these species hull seeds prior to consuming them (Taylor, 1996).

Environmental conditions

Environmental conditions must be monitored continuously to ensure temperature, light, relative humidity, ventilation, and noise stay within ranges appropriate for the species and as specified by the guidance documents. An alarm or alert system should notify facilities personnel when conditions fall out of range, and a plan should be in place to handle environmental control failures that could jeopardize animal health and welfare. Outdoor enclosures, where environmental conditions may widely fluctuate, should include shelter from precipitation and temperature extremes.

Temperature

The recommended dry bulb temperature for poultry in a laboratory setting is 61–81°F (16–27°C) (NRC, 2011). It is important to recognize that the temperature required by chicks can be very different than adults and that a temperature gradient may be necessary to meet physiologic processes (NRC, 2011). Day old poultry chicks, for example, require warmer temperatures than adults, and ambient temperature must be gradually reduced as chicks grow. Domestic fowl and turkey chicks are often housed at a brooder temperature of 35°C at birth with the temperature lowered 3°C every week until the third week of life (Hawkins et al., 2001). Common laboratory psittacines and passerines (e.g., zebra finches and starlings) can tolerate a wide variety of temperature ranges in the wild. Indoor finches do well at a temperature range of 68–77°F (20–25°C) (Nagar and Law, 2010). For species without published temperature guidelines, it is best to research the natural habitat and replicate it as closely as possible.

Light

The photoperiod plays a crucial role in avian physiologic, reproductive, and behavioral health. Poultry chicks, for example, will not eat or drink in low light, and starlings are triggered to begin molting and breeding by changes in day length (NRC, 2011). It is important to have a thorough understanding of the effects of photoperiod on the species at hand and to adjust the light cycle based on life stage, time of year, and research goals. Light timers that allow a gradual shift in light levels to mimic sunrise and sunset are important to use with flight species housed indoors to allow birds to find a roosting spot in the evening and reduce startling that might occur with abrupt changes. Rooms can be outfitted with a dimmer and timer or, alternatively, a dim nightlight. Light quality may also impact avian welfare. Birds can perceive flicker from low frequency florescent tubes (100–120 Hz) and computer monitors that are imperceptible to the human eye (Bateson and Feenders, 2010). It is likely that this flicker is aversive and may also impact behavior (Nagar and Law, 2010). Additionally, birds have retinal cones that allow them to perceive light in the ultraviolet (UV) spectral range. UV light may provide important visual information and has been shown to affect mate choice in some passerines (Bateson and Feenders, 2010). For these reasons, natural light or high frequency broad spectrum florescent light is recommended. A light intensity of 500 lux is generally sufficient for human and animal comfort (Nagar and Law, 2010).

Humidity

There is little data to suggest optimal humidity conditions for avian species. Minor fluctuations are likely of little consequence; however, major changes may serve as an environmental cue in some species. For example, zebra finches, which are opportunistic breeders, will begin breeding after the rainy season begins and seeds sprout (Bateson and Feenders, 2010). Humidity controlled in the range of 30%–70% as generally recommended in The Guide is likely sufficient for the most commonly housed laboratory species (NRC, 2011).

Ventilation

Ventilation serves the purpose of ensuring adequate air quality and affects other environmental conditions such as heat and moisture. Inadequate ventilation in housing areas will lead to heat accumulation, drafts, the accumulation of gases such as carbon dioxide and ammonia, and unacceptable levels of dust. This increases susceptibility to respiratory pathogens and disease in addition to inducing discomfort. The Guide recommends 10–15 air changes per hour in an animal housing room, but this does not account for species differences, heat loads, housing density, type of enclosure, local climate, etc. (NRC, 2011). In general, 15 air changes per hour will be sufficient for most birds housed in laboratory environments to remove dust, CO_2, and ammonia. This rate may need to be adjusted based on the variables identified above.

Noise

Most birds can hear sounds in the frequency range of 1–5 Hz (Bateson and Feenders, 2010). Humans, in contrast, hear sounds best in the range of 0.2–8 Hz (Bateson and Feenders, 2010). Ultrasound, which has a frequency above the limit of human hearing, is therefore unlikely to cause distress in birds. Some birds such as pigeons, however, are thought to hear sounds below 1 Hz and may be bothered by equipment emitting very low frequency noise (Heffner and Heffner, 2007). Conversely, an extremely quiet environment may distress psittacines since the absence of noise in the wild indicates the approach of a predator (Kalmar et al., 2010). Low background music can alleviate this distress. Passerines and psittacines tend to be quite noisy and should be housed in rooms away from species that prefer quieter environments.

environmental enrichment

Materials and methods that facilitate the expression of species-specific behaviors and promote psychological well-being are generally termed environmental enrichment. Enrichment may be the allowance for interaction with conspecifics, opportunities to exercise, the provision of objects that are manipulated and create cognitive challenges, or combinations of all. Many species are cognitively advanced and benefit from a complex and stimulating environment. Parrots, for example, are capable of mimicking human voices and exhibit lifelong learning (Kalmar et al., 2010). Tool use has been documented in many species including finches, ravens, and thrushes (Hawkins et al., 2003). Parrots, pigeons, and domestic fowl can mentally represent hidden movement such that objects can be located after they have been moved (Hawkins et al., 2003). It has been suggested that for many birds, the provision of a complex environment may be more important than cage size alone (Bateson and Feenders, 2010). Not all enrichment, however, may be received positively. Many avian species are naturally neophobic so frequent introduction of novel objects may act as a psychological stressor. It is therefore important that enrichment programs are developed with consideration of the natural behavior and physiology of the species, and regularly evaluated to ensure that intended goals are reached and negative consequences are avoided.

Social Enrichment

Many avian species are highly social and benefit from group housing. However, it is always important to research and understand the normal social dynamics of the species under consideration since some birds may be prone to aggression at certain times of the year (e.g., breeding season) when housed in single sex or mixed sex groups, or if they are solitary in the wild. When research needs dictate single housing in a social species, birds should still be kept in auditory and visual proximity of conspecifics. In some cases, it may be possible to house a bird singly for the period of data collection and then return the bird to a pair or group at the end of the trial. The benefits of this must be balanced with the potential to disrupt group dynamics in species with stable hierarchies. Some species, specifically psittacines, often enjoy human company and may benefit from regular interaction with human caretakers.

Inanimate Enrichment

Birds are highly motivated to engage in behaviors such as foraging, nest building, perching, and bathing. Environments that limit the expression of these behaviors may be a source of frustration and result in feather pecking and stereotypic behaviors (Hawkins et al., 2001). Inanimate environmental stimuli, such as toys, nest boxes, dust or water baths, and perches allow greater use of the space available, promote engagement in normal behavior, allow cognitive skills to be practiced, and may provide areas for individuals to escape or hide. Some food items are highly preferred by birds and are used to provide both variety in the diet and essential nutrients. Although the following items are often classified as enrichment, they should really be considered the foundation of a well-rounded husbandry program:

- *Forage*: Birds have high energy requirements and spend a significant amount of resources searching for food. Domestic fowl, Japanese quail, and pigeons spend the majority of their time foraging (pecking, scratching, and probing) for food on the ground, as do passerine ground feeders such as finches and starlings (Hawkins et al., 2001). Wild psittacines are reported to spend 4–6 h per day foraging (Kalmar et al., 2010). The provision of readily accessible food in a captive setting limits foraging behavior, which may increase boredom and the appearance of undesirable behaviors. Fortunately, increasing foraging opportunities is a relatively easy way of providing environmental stimulation. Foraging behavior can be encouraged and made more challenging by scattering food in substrate such as bedding (e.g., wood chips, sand, straw, gravel, etc.) or by offering objects that encourage oral manipulation (e.g., bunches of string, food puzzles, etc.). If birds are not housed on solid floors, smaller pans of bedding or crumpled paper, or pieces of turf or litter box mats can be added to the enclosure for birds to forage on (Turner, 2010).

- *Perches*: Perches are important for maintaining foot and muscle health in passerines and some ground-dwelling species. Wild finches and starlings spend much time perching in trees, and domestic fowl and pigeons will perch for roosting and resting. In general, perches should be provided in varying thicknesses and textures for optimal foot health. Untreated natural branches not only meet these criteria but also have some natural springiness that helps maintain agility. Some species will also use perches to bill wipe (e.g., starlings) or chew (e.g., psittacines), thus perches should be made of nontoxic material that is soft enough to chew. Manzanita wood, although popular in lab animal facilities, is quite slippery when wet, is too hard for psittacines to chew, and may induce foot problems (Kalmar et al., 2010). In lieu of natural branches, wooden dowel rods are commonly used, although they may be too hard for some species, lack the springiness of branches, and must be provided in various thicknesses. Sandpaper-coated perches are sometimes used to keep nails trim but may be associated with foot lesions in some species; this is more commonly observed in overweight birds or those with long-term housing. Perch position in the enclosure is an important consideration. In general, birds that perch in trees prefer perches above the shoulder height of human caretakers, whereas a ground dweller prefers perches closer to the ground but high enough to prevent other birds from pecking the feathers of perching birds. Psittacines climb and may do well with a highly placed perch to which they can climb by using bars on the enclosure walls (Hawkins et al., 2001). Finches, in contrast, fly to successively closer perches as they approach food sources so staggered perches at various heights are recommended (Nagar and Law, 2010). Japanese quail prefer perches at a variety of levels so as to establish territories. Adequate perch space should be provided to allow all birds to perch comfortably at once, and for fowl to perch at the same height. Perches should be placed so as not to crowd cages and to maximize flight distance in birds that spend a significant amount of time flying. Additionally, perches should not be placed directly over food or water bowls to prevent soiling with droppings. Perches should be sized so that the bird can comfortably grip without its nails damaging its footpads. Most perches are round although poultry may fare better on perches with flattened tops; however, this is a subject of some debate (Hawkins et al., 2001; Pickel et al., 2010).

- *Nest boxes*: Nest boxes may be the single most important inanimate enrichment item to provide for domestic laying hens (FASS, 2010). Many types of nest boxes are commercially available (FASS, 2010) but should be large enough for the hen to turn around. Plastic tubs or crates are also useful as nest boxes. Hens seem to prefer enclosed to open nesting sites although this is not essential (FASS, 2010). Suitable substrate includes straw, wood shavings, or turf. Pigeons will build rudimentary nests in nest pans using small twigs and straw. Finches are prolific breeders in captivity and will readily use nesting material year round. Nests appropriate for finches include plastic bowls, wicker baskets, and wooden or plastic enclosed nest boxes (Figure 2.6). Material such as sisal string, burlap, cotton, wool, and coconut fiber are good nesting substrates. It is helpful to have nest boxes with tops that can be opened from the outside to facilitate record keeping of newly laid clutches without disturbing the nest.

- *Baths*: Dust or water bathing is necessary for feather maintenance by dispersing lipids, and birds appear to enjoy this activity. Domestic fowl will readily dust bathe given appropriate substrate and seem to prefer sand over litter for this activity (FASS, 2010). Fowl housed on bedding will use dust in the litter to bathe. In cages, boxes with sand, or sawdust,

Fig. 2.6 Plastic finch nest box. The box is mounted on the outside of the cage for demonstration.

can provide opportunity for this behavior although birds may lay eggs in boxes. Pigeons, passerines, and psittacines enjoy water bathing and may bathe in drinking water if not given other opportunities to bathe otherwise. Water can be provided in a bath or by spraying birds with a fine mist using a commercial sprayer. Finches and other small passerines require baths for several hours once a week to keep feathers in good health. Water depth should be no more than 0.5–1 cm to keep birds from accidentally drowning. A flat tray with shallow water can be used for pigeons. Pigeons splash considerably when bathing, so water baths can be placed on larger trays to reduce soaking the floor. Aquatic species require a pool large enough to accommodate swimming. Plastic tubs or childrens' swimming pools can be used for this purpose.

- *Toys/Manipulada*: Many species, but psittacines in particular, will benefit from toys and manipulada such as bells, chains, mirrors, and puzzles. Early exposure to novel toys has been reported to reduce fearfulness later in life (Fox and Millam, 2004). Adaptations can be made to the cage such as swinging ladders and cotton ropes wrapped around spiral rings, or the addition of toys that can be manipulated with the beak and feet. Plastic toys that can be easily broken and that may cause GI obstruction if eaten should be avoided (Kalmar et al., 2010). Pigeons benefit from objects hung from the top or front of the cage, such as mirrors, balls, and bird bells. It is important to note that frequent rotation of toys, common in nonavian species, may induce stress in neophobic birds, or in birds that develop strong preferences for certain objects (Hawkins et al., 2001; Kalmar et al., 2010).

- *Shelter*: In outdoor aviaries or flight cages, the addition of protective cover such as shrubs and other plants serves to increase the complexity of the environment, provide natural perching spots, and provide protective cover from humans, predators, and conspecifics. The additional visual barriers can also help reduce anxiety.

- *Food*: Preferred food items, such as invertebrates, fruit, or greens, provide variety to the daily diet in addition to serving as a source of essential vitamins and minerals. Offering food in the "natural state," such as millet sprays and unshelled nuts, can encourage foraging behavior. Additionally, food can be placed in a puzzle feeder or on a foraging board to provide additional mental stimulation. Some birds can be trained to perform operant tasks such as pecking a keyboard or hopping on a perch in return for a food reward.

Conditions Secondary to Poor Housing Environments

Feather pecking

Feather pecking in poultry and waterfowl consists of either nontraumatic pecking at the feathers of other birds or more aggressive pecking that results in feather loss and injuries (FASS, 2010). In extreme cases, birds will cannibalistically pick skin and underlying tissues around the tails, vent, toes, and wings, leading to severe injury and death. Feather pecking in fowl is a poorly understood behavior but appears to be related to misdirected foraging behavior (Blokhuis and Wiepkema, 1998). Other factors include increased stocking density, bright light, and genetic background (Blokhuis and Wiepkema, 1998; Hawkins et al., 2003). Feather pecking is difficult to control once started because birds are attracted to damaged feathers, so prevention and early intervention is best (FASS, 2010). Provision of foraging substrate (straw, pecking blocks, string, and litter) and/or reducing group size are the first steps taken to reduce feather pecking. Temporarily lowering the light intensity or using red light is also reported to reduce the incidence in some cases (Blokhuis and Wiepkema, 1998; Hawkins et al., 2003). Physical modifications used to limit the ability of birds to induce physical harm, such as beak trimming, may be necessary in situations where feather picking and cannibalism remain a problem in spite of other control measures.

This practice is controversial. Only experienced personnel should perform beak trimming, and chicks should be younger than 10 days of age (FASS, 2010). In psittacines, feather picking is self-directed; however, similar to feather pecking, the behavior can be difficult to address once started. Evaluation of the enrichment program to ensure the environment is adequately complex, and not unintentionally eliciting fear, should be considered first. In severe cases, pharmacologic treatment may be required.

Stereotypic behavior

Stereotypic behaviors are repetitive patterns of movement that appear to have no obvious function and are considered to be manifestations of chronic distress and frustration due to physiological or psychological stressors (Garner, 2005). Examples of common stereotypies in birds are pacing, circling, pecking at one spot, and repetitive manipulation of an object. Allowing opportunities for birds to engage in species-specific behaviors, such as foraging, bathing, and interacting with conspecifics, has been shown to reduce stereotypic behavior in birds (Meehan et al., 2004; FASS, 2010). In some cases, however, even after the underlying cause has been identified and removed, the undesirable behavior may continue. This may be more reflective of the bird's history than the current housing conditions (Kalmar et al., 2010).

Vocalization

Psittacines have a wide vocal repertoire and will vocalize frequently and loudly under normal conditions. In some cases, vocalization may become compulsive or excessively loud. Although excessive loud vocalizations may represent true distress, they may also arise from inappropriate, learned attention-seeking behavior, and thus it is important to distinguish between the two (Kalmar et al., 2010).

Disease

Foot problems can arise from inappropriately sized perches, improper flooring material, or wet bedding. Both perches soiled with feces and wet litter are associated with an increase in foot dermatitis. Perches that are too small may not allow proper gripping such that the bird's toenails damage the footpad, whereas perches that are too large may not allow birds to properly grip. Birds that lack access to dust and water bathes will show poor plumage. Excessive ammonia levels may predispose birds to respiratory disease. Poor nutrition is associated with many avian conditions such as osteoporosis and poor feather growth.

Sanitation

The frequency and method of cleaning and disinfecting rooms, enclosures, and accessories will depend on the species housed, the stocking density, the type and size of enclosure, the life stage of the animal, and the environmental conditions. There is no one-size-fits-all program, and most will be based on professional judgment to a large extent and recommendations from The Guide. At a minimum, the schedule of sanitization should be frequent enough to keep birds clean, dry, comfortable, and healthy, and to minimize accumulation of ammonia. General recommendations and considerations include the following:

- Many birds are easily stressed by the commotion created by daily husbandry procedures. It is best for husbandry duties to be quietly carried out at the same time each day by a consistent, familiar person. Additionally, cage designs that reduce the need for personnel to enter or disturb cages, for example cage bottoms with sliding drawers, can decrease negative effects associated with husbandry activities.

- All room surfaces and equipment should be thoroughly cleaned and sanitized before a new group of birds arrives or when a room is depopulated. It is especially important that newly hatched chicks go into a clean environment.

- Floors should be swept and surfaces wiped regularly to remove debris and to prevent accumulation of feather dust. Feather dust may also accumulate on cages and must be removed regularly with a damp cloth. A low noise, high-efficiency particulate arrestance filtered vacuum can be useful for picking up small debris such as seeds while minimizing dispersal of fine particulates into the air.

- Dropping collectors/trays and paper liners should be cleaned 1–3 times per week or more frequently if needed. Small cages, cages with wet flush systems, or cages housing species that produce a large amount of droppings (e.g., starlings) may require daily cleaning.

- Bedded floor pens should have wet litter and droppings removed daily or as often as necessary to keep ammonia levels acceptable and birds clean and dry. Litter can be topped off with fresh substrate. A complete bedding change will be necessary every few months. If housing breeding birds, the schedule of cleaning must be carefully timed so as not to disrupt egg laying.

- The cage sanitation schedule will depend on the type of cage and the stocking density. Individual cages or cage racks are typically cleaned every 2 weeks. Most commercially manufactured cages are designed to withstand mechanical cage washers that clean using a combination of detergents and hot rinse water at 180°F to disinfect. Larger flight cages are often sanitized every few months or when excessively soiled or depopulated via chemical disinfection (e.g., quaternary compounds, peroxygen compounds, diluted bleach 1:10, etc.) and a thorough rinse.

- Feeders and waters should be cleaned 1–2 times weekly or more often if visibly soiled.

- Perches and other accessories should be washed and disinfected as they become soiled.

- Food bowls, water bottles, and other accessories that are too small or delicate to withstand mechanical cage washers can be washed by hand. At the authors' institution, these items are washed first with a mild dish detergent to remove gross debris, then soaked in a 1:10 bleach solution for 5 min followed by a clean water rinse.

- The amount of adenosine triphosphate (ATP) on a surface is a measure of organic debris. To assess sanitation effectiveness, ATP levels can be measured using portable, commercially available units that are easy to use. In general, a swab provided is wiped on the surface of interest and placed into a handheld device that utilizes a bioluminescence reaction to quantify the level of ATP. The level is reported as light units. The absence of ATP indicates that organic debris has been removed from the surface. Acceptable or "passing" ATP levels are determined by the institution based on sanitation program goals. The reader is referred to the CDC Guideline for Disinfection and Sterilization of Healthcare Facilities, 2008 for an overview of approaches, factors affecting, and agents used for disinfection and sterilization (Rutala and Weber, 2008).

Pest control

A program to exclude and eliminate vermin, such a cockroaches, flies, mosquitoes, wild rodents, and, if housed outdoors, predators and wild birds, should be instituted to control feed contamination, disease transmission, stress, and loss of birds. Rodents in particular may harbor diseases which might transfer to birds. Any openings and cracks should be sealed to prevent building entry, and vermin breeding sites should be eliminated. Some pesticides may be toxic to birds and have a deleterious effect on research data; hence, their use should be minimized and carefully considered in consultation with veterinary and research staff. When developing a comprehensive program, it is often helpful to contract with an outside pest control vendor with experience in working with animal research facilities.

Identification and Record Keeping

Identification

Identification of birds may be necessary for research or regulatory purposes and is generally a valuable component of colony management. Individual identification methods may be permanent or temporary, and not all methods are appropriate for all avian species. In some situations it may neither be practical nor necessary to mark individual birds, in which case identification by the enclosure or group is sufficient. Choosing the least invasive method that will meet experimental and management goals is always best.

Methods

- *Documentation of physical or behavioral differences*: Plumage patterns and colors, morphologic differences, song, etc. may be used to identify individuals or gender groups. Starlings, zebra finches, Japanese quail, and pigeons have sexually dimorphic features that allow gender differentiation.

- *Marking or dyeing feathers*: Nontoxic dye can be applied to the feathers as a temporary identification method that will be lost when the bird molts. Some colors may impact the behavior of conspecifics; therefore, potential effects should be thoroughly researched beforehand (Hawkins et al., 2001).

- *Leg rings/bands*: One or more colored, numbered, plastic or metal rings can be applied to the leg of the bird (Figure 2.7). Rings are often placed in various color combinations to aid in visual identification without the need for handling. Rings may be closed or split, and must be fitted for the age and species of bird so that the rings can move freely on the leg without falling off. Split rings can be applied at any age using a special applicator tool, whereas closed rings are slid over the foot onto the leg and are only appropriate for mature birds. When using rings in growing animals, birds must be checked

Fig. 2.7 Plastic, colored, numbered split ring leg bands were placed on this finch for identification.

frequently to ensure the ring is not becoming constrictive. Colored rings may impact behavior in some species (Hawkins et al., 2001). Metal bird bands used for tagging wild birds in field research are issued by the United States and Canadian government and require a special permit. Readers are referred to the Bird Banding Laboratory (http://www.pwrc.usgs.gov/bbl/) for more information.

- *Wing tags*: Wing tags are placed through the web (patagium) on the front edge of the wing, taking care to avoid the musculature. The tags are labeled with number or letter combinations and come in a variety of colors. Many species tolerate tags well; however, those capable of flying should be monitored to ensure the tags do not impede flight. Rodent ear tags can be used for very small birds.

- *Electronic tags (microchip)*: A small chip containing an electronically read code can be implanted in the pectoral muscle or alternatively in the subcutaneous tissue at the base of the neck. This method is generally not appropriate for birds weighing less than 100 g (Granzow, 2008).

Record keeping

Accurate record keeping is an essential component of colony management. Regular review of records can help identify colony health issues related to disease, experimental manipulations, or husbandry procedures. Records may pertain to individuals or groups of birds and should contain clear, consistent, and readily retrievable information.

- *Individual records*: Typical individual records contain species, animal identification, source, birth or acquisition date, sex, exit date, and final disposition. The experimental history and/or medical information may also be contained in individual records.

- *Group records*: Group records are used when individual animals are not identified, or when tracking group-related activity, such as experimental schedules, social housing, and census.

- *Medical records*: Medical records contain clinical and diagnostic information such as observations, preventive care, treatments, experimental manipulations, surgical procedures, and laboratory findings. These may be organized by the individual or the group.

- *Breeding records*: Breeding records are useful for tracking reproductive performance. Records may include identification of breeding pairs, number and date of eggs laid, hatch dates, sex of offspring, and numbers of successfully reared chicks.

- *Mortality records*: Mortality records document the identity, location, date of death, and cause of death if known. An analysis of trends can provide important information about disease prevalence or contributing factors such as husbandry or experimental procedures.

Transportation

Transport is considered a significant stressor to birds. Minimizing stress to ensure welfare and successful movement is therefore a primary objective. To that end, the major considerations include the design, size and construction of the transport container, the provision of food and water, protection from temperature extremes, and reducing the exposure to excessive auditory and visual stimuli that might startle the bird. All of these basic tenets apply regardless of whether birds are transported just a short distance within an institution or shipped to another state.

Birds used in research are currently not defined as animals in the United States Department of Agriculture (USDA) Animal Welfare Act and Regulations (AWAR). As such, there are no requirements for birds promulgated by the USDA where transport is concerned. This being the case, experience and professional judgment are often instrumental in the development of shipping procedures. Fortunately, there are

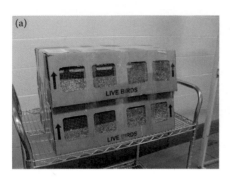

Fig. 2.8 (a) Avian shipping crate. (b) A moistened sponge can be used as a water source for small passerines.

also several publications that offer guidance and, in some situations, the legal requirements for the transport of avian species.

The most comprehensive document available is the *Live Animal Regulations* published by the International Air Transport Association (IATA), in its 41st edition as of this writing. While this document is intended to address the shipment of animals via airplane, Section CR 11-23 contains detailed descriptions of container construction and design, stocking densities, and general guidelines for food and water provision during transport for hundreds of avian species. This information may be useful in any situation where avian species must be moved from a field collection site to the vivarium. Some of the requirements that apply to all species as written in Section CR 11-23 include the following:

- Birds will be carried only in closed containers (Figure 2.8).
- Containers must not be constructed of materials containing chemicals or toxins, for example, chemically impregnated wood.
- Wooden perches must be provided for birds that rest by perching, and there must be sufficient space for each bird to perch together and enough height to allow perching with its head upright and tail clear of the floor.
- The container must be clean and leak-proof with bedding that is suitable for the species.
- The container must be adequately ventilated on at least three sides with ventilation openings small enough or baffled to prevent egress of the bird.
- Separate food and water troughs must be provided, either fixed or attached so they are accessible for replenishment.

Beyond container specifications, the IATA publication contains a wealth of information about commercial carrier requirements and permits necessary for importation. There are additionally two resources that the reader might find useful:

The Guide for the Care and Use of Agricultural Animals in Research and Teaching (FASS, 2010) provides some guidelines for the transport of poultry.

The Guidelines to the Use of Wild Birds in Research published by the Ornithological Council (Fair et al., 2010) discusses general considerations for avian transport.

references

Bateson, M., and Feenders, G. 2010. The use of passerine bird species in laboratory research: Implications of basic biology for husbandry and welfare. *ILAR.* 51:394–408.

Blokhuis, H.J., and Wiepkema, P.R. 1998. Studies of feather pecking in poultry. *Veterinary Quarterly.* 20:6–9.

CEC. 2006. Appendix A of the European Convention for the Protection of Vertebrate Animals Used for Experimental and Other Scientific Purposes. Guidelines for the Accommodation and Care of Animals. Council of Europe, Strasburg.

Dawkins, M.S. 1983. Cage size and flooring preference in litter-reared and cage-reared hens. *British Poultry Science.* 24:177–182.

Fair, J., Paul, E., and Jones, J. (eds.). 2010. Guidelines to the Use of Wild Birds in Research, Washington, D.C.: Ornithological Council. Online: http://www.nmnh.si.edu/BIRDNET/guide/index.html.

Federation of Animal Science Societies (FASS). 2010. *The Guide for the Care and Use of Agricultural Animals in Research and Teaching*, 3rd edition. Champaign, IL: FASS. Online: http://www.fass.org.

Fox, R.A., and Millam, J.R. 2004. The effect of early environment on neophobia in orange-winged Amazon parrots (*Amazona amazonica*). *Applied Animal Behaviour Science.* 89:117–129.

Garner, J.P. 2005. Stereotypies and other abnormal repetitive behaviors: Potential impact on validity, reliability, and replicability of scientific outcomes. *ILAR.* 46:106–117.

Granzow, E. 2008. Microchip placement for identification of birds. *Laboratory Animals.* 37:21–22.

Hawkins, P., Bairlein, F., Duncan, I., Fluegge, C., Francis, R., Geller, J., Keeling, L., and Sherwin, C. 2003. Future principles for housing and care of laboratory birds. Report for the revision of the Council of Europe Convention ETS123 Appendix A for birds. Online: http://www.coe.int/t/e/legal_affairs/legal_co-operation/biological_safety_and_use_of_animals/laboratory_animals/GT_123_2003_6rev_Birds.pdf.

Hawkins, P., Morton, D.B., Cameron, D., Cuthill, I.C., Francis, R., Freire, R., and Townsend, P. 2001. Laboratory birds: Refinements in husbandry and procedures. Fifth report of BVAAWF/FRAME/RSPCA/UFAW joint working group on refinement. *Laboratory Animals*. 35:1–163.

Heffner, H.E., and Heffner, R.S. 2007. Hearing ranges of laboratory animals. *JAALAS*. 46:11–13.

Howard, B.R. 1992. Health risks of housing small psittacines in galvanized wiremesh cages. *JAVMA*. 200:1667–1674.

Kalmar, I.D., Janssens, G.P.J., and Moons, P.H. 2010. Guidelines and ethical considerations for housing and management of psittacine birds used in research. *ILAR*. 51:409–423.

Koutsos, E.A., Matson, K.D., and Klasing, K.C. 2001. Nutrition of Birds in the order psittaciformes: A review. *Journal of Avian Medicine and Surgery*. 15:257–275.

Lay, D.C., Fulton, R.M., Hester, P.Y., Karcher D.M., Kjaer, J.B., Mench, J.A., Mullens, B.A. et al. 2011. Hen welfare in different housing systems. *Poultry Science*. 90:278–294.

Meehan, C.L., Garner, J.P., and Mench, J.A. 2004 Environmental enrichment and development of cage stereotypy in Orange-winged Amazon parrots (*Amazona amazonica*). *Developmental Psychobiology*. 44:209–218.

Nagar, R.G., and Law, G. 2010. The zebra finch. In *The UFAW Handbook on the Care and Management of Laboratory Animals*, 8th edition, edited by Hubrecht, R., and Kirkwood, J. Oxford: Wiley-Blackwell.

NRC. 2011. *Guide for the Care and Use of Laboratory Animals*, 8th edition. The National Academies Press, Washington, DC.

Pickel, R., Schrader, L., and Scholz, B. 2010. Pressure load on keel bone and foot pads in perching laying hens in relation to perch design. *Poultry Science*. 90:715–724.

Pohle, K., and Cheng, H.W. 2009. Furnished cage system and hen well-being: Comparative effects of furnished cages and battery cages on behavioral exhibitions in white leghorn chickens. *Poultry Science.* 88:1559–1564.

Rutala, W.A., and Weber, D.J., Healthcare Infection Control Practices Advisory Committee. 2008. Guideline for Disinfection and Sterilization in Healthcare Facilities, 2008. Online: http://www.cdc.gov/hicpac/pdf/guidelines/Disinfection_Nov_2008.pdf.

Schulte, M.S., and Rupley, A.E. 2004. Avian care and husbandry. *Veterinary Clinics of North America: Exotic Animal Practice.* 7:315–350.

Taylor, E.J. 1996. An evaluation of the importance of insoluble vs. soluble grit in the diet of canaries. *Journal of Avian Medicine and Surgery.* 10:248–251.

Turner, T. 2010. Enrichment for Carneaux pigeons used in behavioral learning research. *Laboratory Animals.* 39:40–41.

management

---

regulatory agencies and compliance

The applicability of existing federal and state policies and regulations as well as oversight by regulatory bodies varies by species used and research situation. In the United States, governmental oversight of the use of animals in biomedical research is provided by the United States Department of Agriculture (USDA) through its Animal Welfare Act (AWA) and Regulations (AWR) and the National Institutes of Health (NIH) through the Public Health Service (PHS) Policy on Humane Care and Use of Laboratory Animals (aka PHS Policy), a product of the Health Research Extension Act (HREA) of 1985.

The USDA AWR do not include birds in its definition of animals, stating that the term "animals" excludes birds, rats of the genus *Rattus*, and mice of the genus *Mus* bred for use in research. The definition also excludes poultry used or intended for use as food or fiber or intended for improving animal nutrition, breeding, management, or production efficiency, or for improving the quality of food or fiber. This being the case, the care and use of birds in research is not the concern of the USDA and hence provides no standards for housing and husbandry.

In contrast to the AWR, PHS policy defines an animal as any live, vertebrate animal used or intended for use in research, research training, experimentation, or biological testing or for related purposes. The use of birds must therefore be in accordance with the PHS policy and the tenets of the HREA if the institution where the work takes place receives PHS funds (e.g., from NIH). The Office of Laboratory Animal Welfare (OLAW), which resides in the NIH Office of Extramural Research, is generally responsible for administration and interpretation of PHS policy. OLAW maintains a website (https://grants.nih.gov/grants/olaw/olaw.htm) with substantial information, including frequently asked questions about PHS policy interpretation and implementation. With respect specifically to birds, one FAQ does state that PHS policy applies to eggs only after hatching. PHS policy otherwise contains no specific language pertaining to birds, but instead refers to *The Guide for the Care and Use of Laboratory Animals* (aka, The Guide) for housing and care recommendations.

For studies in the field where birds or their parts (e.g., feathers), nests, or eggs are collected, a Migratory Bird Scientific Collecting permit issued by the United States Fish and Wildlife Service (USFWS) is necessary. State permits might also be required in order for the federal permit to be valid. It should also be noted that PHS policy extends to field studies if they are supported by the PHS funds. For studies that occur internationally, rules and regulations put forth as part of the Convention on International Trade in Endangered Species of Fauna and Flora (CITES) might become applicable. The reader is referred to an excellent publication of the Ornithological Council (Paul, 2005), and to the federal registry, 50 CFR Part 14, Subpart J, *Standards for the Humane and Healthful Transport of Wild Mammals and Birds to the United States,* for the laws and guidance in all matters of live bird importation. Regardless of the nature of a field study, it is best to consult with local, state, and federal authorities to determine what permits are necessary for collecting.

institutional animal care and use committee

An Institutional Animal Care and Use Committee (IACUC) is charged with providing oversight on behalf of the institution of all aspects of the care and use of animals used for teaching and research. The IACUC office ensures compliance with all applicable federal regulations through adherence to The Guide, AWR, and PHS policy. The *Guide for the Care and Use of Agricultural Animals in Research and Teaching* (aka, The Ag Guide) is also used as a standard for animals housed in an agricultural setting, for example, where poultry are kept to study aspects of meat or egg production. Additionally, policies and position statements promulgated by OLAW and the Association for the Assessment and Accreditation of Laboratory Animal Care (AAALAC) International are helpful for institutions that receive PHS funds and are accredited by AAALAC. Where no other guidance exists for a particular species, publications from professional organizations such as the Ornithological Council can be useful in developing institutional policies.

The AWA and PHS policy state the requirements for the composition and function of IACUCs. Perhaps, the most important functions of an IACUC are conducting the semi-annual inspections of animal housing facilities and laboratory spaces and the review of programs for animal care at the institution. In order for these to be meaningful, IACUC members engaged in these functions need to be reasonably well versed in the care and use of the avian species housed. It is advisable for an IACUC office to provide training and resources to ensure that inspectors are effective in their duties. This training should address housing standards, general health, disease recognition and management, and any additional issues specific to the institution.

Another important IACUC function is the review of research protocols involving animals. Veterinarians and scientists serving on the IACUC will need to develop familiarity with standards of care and proper procedures where avian species are used. Appropriate anesthesia and analgesia, surgical technique, tissue collection, handling, and proper euthanasia are just a few aspects that require critical evaluation. Other chapters in the present work provide information helpful in most of these areas. Euthanasia methods must adhere to the standards of the most current AVMA Guidelines for the Euthanasia of Animals.

sources of birds and procurement

At most institutions, birds will be obtained from either commercial vendors or from the wild. The source is highly dependent upon the species required and goal of the study. There are several vendors of poultry species, for example, but few, if any, for some of the songbird species that are routinely used in research. There will, therefore, be occasions when pet store suppliers are the only source available. Regardless of the origin, it is important that the receiving institution critically evaluates a vendor's health management program and determines whether the vendor provides a quality product. Subsequent to that, appropriate receiving and quarantine procedures should be implemented as deemed necessary by professional judgment.

Procuring birds from vendors is usually a well-defined process with animal care programs and fairly easy to manage. If species are captured in the field and brought to a facility for housing, the typical ordering process is bypassed, requiring good communication between research and animal care staff to accommodate animals upon arrival. As mentioned, in studies where birds are captured in the field, the USFWS and state wildlife agencies would have purview and issue appropriate permits to allow collection of specimens. The IACUC would typically verify with the researcher that these are in place before work begins.

quarantine and conditioning

Quarantine programs serve as a vital mechanism to reduce the risk that the resident population is exposed to pathogens that might be unknowingly brought into the facility with imported animals. The specific practices that are employed as part of the program are largely dictated by the source of the animals. Birds coming from reputable sources with rigorous health monitoring programs might require a little to no quarantine period, whereas wild caught specimens should always undergo appropriate evaluation and necessary treatment prior to introduction to the resident census. Preemptive treatment of the latter specimens for subclinical conditions such as endoparasitic infestation should be considered as part of the quarantine process.

When a defined quarantine period is deemed unnecessary, at a minimum, a period of conditioning or acclimation for animals recently imported into a facility should be observed. Shipping and/or capture in the case of wild-caught animals, is typically the source of extreme stress in birds. Small passerines in particular can succumb to the stress response associated with these events and formerly subclinical conditions can manifest. IACUCs at most institutions have policies mandating that experimentation not begin until animals have been in the facility for a prescribed period of time, typically 3–7 days. Such policies are important not only to the health of the animals, but also to the integrity of the science as this acclimation time reduces the risk that shipping stress will influence experimental outcomes.

occupational health and safety considerations

At institutions where work with birds is conducted, occupational health and safety program (OHSP) professionals need to become familiar with hazards that are inherent to the work. The types of hazards that the OHSP must assess are dependent largely on the species involved and the type of manipulations performed. In all cases, it is the responsibility of the OHSP to identify hazards and then perform an assessment of risk to workers with subsequent implementation of proper practices and controls to ensure worker safety.

As with nearly any animal species, the possibility for human injury from a bite or scratch during restraint and handling exists (see Chapter 5 for proper handling techniques). The risk of injury is minimal when working with small passerine species. In contrast, bites from large psittacines can cause significant injury requiring medical attention. Similarly, raptors can inflict bite wounds and severe puncture wounds using their powerful talons. Regardless of the type of bird, proper restraint of the head, feet, and wings significantly mitigates the risk of injury not only to the handler, but also to the bird, making training in proper handling and restraint technique most important. Hazards inherent to the capture or collection of birds in the field are comparable to those associated with field studies involving other species. These include exposure to extreme weather conditions, biting and stinging insects and arachnids, zoonotic disease carried by vectors such as ticks and mosquitoes, venomous reptiles, and injury secondary to exertion in an outdoor setting. If samples are collected from nesting animals, hazards associated with climbing might also come into play. Proper education and training are the most effective tools for reducing the risk of injury in all scenarios.

The potential for zoonotic disease transmission from avian species is relatively low, but full consideration of the risk should be given based upon an assessment of a particular species or situation. Psittacosis (aka "parrot fever") is caused by *Chlamydophila psittaci*, a Gram-negative bacterium. Despite the name, species other than psittacines have been shown to carry the agent (see Chapter 4). The risk to personnel in the research setting is unknown. The Centers for Disease Control and Prevention (CDC) received reports of 813 cases from 1988 to 1998 that were presumed to be acquired from pet birds, primarily psittacines. In light of that, the animal care program should employ personal protective equipment (PPE) and practices to reduce potential exposure should it be deemed appropriate (Smith et al., 2005).

Where field studies and wild-caught birds are concerned, the list of potential zoonotic agents expands considerably. A comprehensive discussion is beyond the scope of this work, but there are a few worth mentioning briefly. *Salmonella* spp. are commonly carried by wild birds, with *Salmonella typhimurium* representing one of the most common isolates (Hughes et al., 2008). *Campylobacter jejuni*, *Yersinia pseudotuberulosis*, and *Y. enterocolotica* can all likewise be carried in wild populations. Proper attention to hand washing during handling and collecting is the most effective way to prevent infection. Newcastle disease caused by Avian paramyxovirus 1 has been identified in over 230 species of birds and presents a small risk to workers during the collection of clinically healthy birds (Dortmans et al., 2011). Avian influenza, or Influenza A, can be carried by several species of passerines. Only the low pathogenicity strains exist in the United States while highly pathogenic H5N1 strains have been isolated in parts of Asia, Africa, and Europe (Chen et al., 2006; Normile, 2006). It should, however, receive some consideration when collecting species in the United States or Canada that might comingle during migration with populations from regions where high pathogenicity strains are endemic. West Nile virus, a Flavivirus, has been isolated in 317 species of birds serving as a reservoir for the virus. Humans are exposed only through bites from mosquitoes of the genus *Culex*, the primary vector for virus transmission, making exposure control through the use of insect repellents most prudent (Gray and Webb, 2014). Histoplasmosis is a fungal infection caused by *Histoplasma capsulatum*. The organism thrives in soil where bird droppings accumulate, and is considered endemic to the Ohio and Mississippi River valley regions. Although there is no risk of disease transmission from the animals, collecting in areas where birds like European starlings (*Sturnus vulgaris*) gather en masse, for example, creates some risk of exposure.

Birds host a wide variety of ectoparasites, none of which are a major risk to humans. The red mite, *Dermanyssus gallinae*, can transiently infest humans, causing a pruritic dermatitis. The other mites, lice, fleas, and ticks that can potentially feed on birds are either rare or host-specific and of little concern. If wild-caught species are housed in a facility, a comprehensive parasite control program should be in place primarily to ensure the health of the animals, but also to mitigate the small risk to humans that does exist.

Finally, exposure to allergens and resultant hypersensitivity in workers should be addressed through the institution's health screening process. Very little regarding human allergy secondary to bird exposure is found in the scientific literature, but a hypersensitivity pneumonitis known colloquially as Bird Fancier's Lung (BFL) or bird breeder's disease is described (Cooper et al., 2014). The most reactive antigens are shed in the feces and some are associated with dander as well. It is inhalation of high concentrations of these substances that is presumably responsible for the pneumonitis. The primary risk to laboratory and husbandry personnel would most likely come from the handling of soiled housing materials such as cage linings during cleaning of enclosures.

---

references

Chen, H., Smith, G.J., Li, K.S., Wang, J., Fan, X.H., Rayner, J.M., Vijaykrishna, D. et al. 2006. Establishment of multiple sublineages of H5N1 influenza virus in Asia: Implications for pandemic control. *Proceedings of the National Academy of Sciences of the United States of America*. 103:2845–2850.

Cooper, C.J., Teleb, M., Elhanafi, S., Ajmal, S., and Hernandez, G.T. 2014. Bird fanciers' lung induced by exposure to duck and goose feathers. *The American Journal of Case Reports*. 15:155–158.

Dortmans, J.C., Koch, G., Rottier, P.J., and Peeters, B.P. 2011. Virulence of Newcastle disease virus: What is known so far? *Veterinary Research*. 42:122.

Gray, T.J., and Webb, C.E. 2014. A review of the epidemiological and clinical aspects of West Nile virus. *International Journal of General Medicine*. 7:193–203.

Hughes, L.A., Shopland, S., Wigley, P., Bradon, H., Leatherbarrow, A.H., Williams, N.J., Bennett, M. et al. 2008. Characterisation of *Salmonella enterica* serotype *typhimurium* isolates from wild birds in northern England from 2005–2006. *BMC Veterinary Research*. 4:4.

Normile, D. 2006. Avian influenza. Evidence points to migratory birds in H5N1 spread. *Science*. 311:1225.

Paul, E. 2005. A guide to the permits and procedures for importing live birds into the United States for scientific research and display. The Ornithological Council, Chevy Chase, MD. Online: http://www.nmnh.si.edu/BIRDNET/documents/Importguidelivebirds.pdf.

Smith, K.A., Bradley, K.K., Stobierski, M.G., and Tengelsen, L.A. 2005. Compendium of measures to control *Chlamydophila psittaci* (formerly *Chlamydia psittaci*) infection among humans (psittacosis) and pet birds, 2005. *Journal of the American Veterinary Medical Association*. 226:532–539.

veterinary care

general physical examination

The avian exam should start without physically handling the animal. The environment and initial observations of the bird will provide critical information about its health status and maximize the benefit of the physical exam. In addition, restraint times should be minimized because birds are easily stressed, and just the stress of handling can cause a very ill bird to die. In severely sick birds, especially in respiratory distress, the animal may benefit more from a quiet, dark, and oxygen-enriched environment prior to handling. However, a brief but thorough physical exam is warranted for the majority of animals even when ill. The veterinarian should watch restrained animals carefully for signs of excessive stress. Respiratory rate will increase in all restrained birds, but closing of the eyes and/or becoming less responsive warrant releasing the animal back into the cage. Body weight measured with a gram scale should be obtained as part of the physical exam whenever possible. Body temperature is not a routine part of a physical exam as it normally increases significantly with handling (Greenacre and Lusby, 2004). Details pertaining to restraint for specific species are described in Chapter 5: Experimental Methodology. A fecal Gram stain (described later) can be a useful diagnostic tool for birds, and should be considered for sick birds where the cause of illness is not immediately evident.

Examination of the Environment

- *Food*: Observe for evidence that the bird has been eating. Birds that are sick may pretend to eat and knock food out of the bowl, but remnants of crushed pellets or seed hulls indicate the food has actually been processed and likely ingested. Decreased fecal production can also indicate inappetence.

- *Feces*: Feces are typically of a well-formed shape within the urates and urine, although the consistency varies between species and is less cohesive in some birds such as chickens. Fecal color is usually a light to dark brown or green color, but the color is influenced by the diet. Birds that eat heavily pigmented fruits such as raspberries or blackberries can have dark stool colors which could be confused with melena. Colored pelleted diets can also modify fecal color. Diarrhea is uncommon in birds except with infectious agents and is characterized by looser stool and occasionally gas bubbles (Figure 4.1). Feces which can appear like diarrhea due to stress and polyuria can occur because of the fear associated with an observer during an exam, but the other feces in the cage should be normal in this case.

- *Urates*: Urates should be a white to slight yellow in color. Green or dark yellow urates can be a symptom of liver disease, resulting from excess biliverdin.

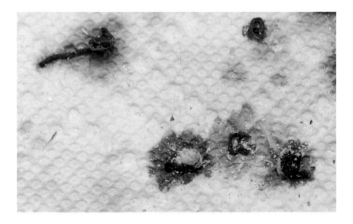

Fig. 4.1 Feces and urine from a normal cockatiel (*Nymphicus hollandicus*) on a diet of colored pellets. The color of the pellets affects the appearance of the feces and urine, which could be confused with the droppings of an animal with liver disease or hematuria.

- *Urine*: Urine should be clear, and the color can be affected like urates in animals with liver disease. The urine can reflect the color of the feces when birds are fed a colored pelleted diet (Figure 4.1).

- *Vomit*: Birds typically shake their heads when they vomit. Remnants of dried vomitus may be found on the bird's head or cage bars.

- *Housing*: Observe caging for any broken objects or other sources of potential harm for the animal. For large birds that chew, observe for any objects that might have missing pieces or damage that suggest the bird may have a foreign body or have been exposed to toxic substances through ingestion.

Animal Observations Prior to Handling

- *Alertness and activity level*: Birds should sit with their eyes open and flat feathers, particularly in the presence of an unfamiliar observer. Although some birds will fluff their feathers if resting when they are cold and may normally explore the bottom of a cage, a bird with fluffed feathers while closing its eyes and sitting in one area is likely sick. Parrots may "flash" as a warning when they are irritated, which involves rapid dilation of their pupils.

- *Respiration*: Observe the animal in the cage for increased respiratory rate or effort before handling, as handling will exacerbate dyspnea. A pronounced movement of the tail with respiration, termed a "tail bob," can indicate increased respiratory effort. Birds do not open mouth breathe unless in severe respiratory distress, although some yawn or pant if stressed or hot. Birds that are hot will also hold their wings away from their body. Because birds do not have a diaphragm, respiratory distress can result from primary respiratory pathology or a space occupying coelomic mass impinging on the respiratory system.

- *General feather condition:* A general assessment of the feather condition and symmetry is easier to do while looking at the unrestrained animal. The feathers should have smooth edges and have a slight sheen. Focal feathers out of place may indicate a feather abnormality or an underlying problem such as a mass or mites (Figure 4.2). A substantial swelling from subcutaneous emphysema associated with a ruptured air sac would likely be obvious without needing to handle the animal.

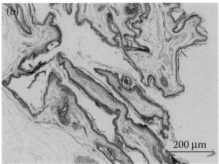

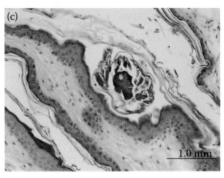

Fig. 4.2 (a) White-throated sparrow (*Zonotrichia albicollis*) that presented for aggression and focal feather abnormalities with associated photomicrographs (b) and (c). (b) Two mites (one centrally located in the image, the second ~1.0 mm below) are present within the stratum corneum of the epidermis. Associated pathological changes include mild irregular epidermal hyperplasia (acanthosis) and mild hyperkeratosis. (c) A high magnification of the previous photomicrograph. The mite is approximately $100\,\mu m$ wide $\times\,300\,\mu m$ long with a chitinous exoskeleton, articular appendages, striated muscle, a gastrointestinal tract, and ovarian tissue with developing eggs. (Photomicrographs courtesy of Jennifer Asher.)

Generalized feather dullness and feathers with uneven edges are abnormal. Feathers are covered by a sheath as they grow (pin feathers), which should fall off or be removed by the bird once the feather growth is complete. Most indoor birds molt intermittently throughout the year. Some birds molt all of their feathers in a short period of time seasonally, but should shed the sheaths within a few days. Birds are normally fastidious groomers, so poor feather condition or an abundance of pin feathers suggests illness secondary to a lack of hygiene.

- *Wing symmetry*: The position of the wings should be observed. A subtle wing droop can be detected by comparing the height of the carpal joint with the other side and can indicate a musculoskeletal injury such as a coracoid fracture.

- *Locomotion:* Many birds will stand still during observations, but can be encouraged to move by slowly opening the cage. Observe equal weight bearing on both legs. For birds housed in a flight cage, observe for normal flight.

Sex Determination

The ability and technique to determine the sex of the bird based on visual exam will depend on the species, and techniques for representative orders are described hereunder. All birds can be differentiated by polymerase chain reaction (PCR) testing. Blood is the ideal tissue for PCR testing but it can be accomplished with other tissues such as feathers. Tissues other than blood have a theoretical risk of giving unreliable results due to contamination in group housed animals, although practically the risk is likely low. Diagnostic laboratories that perform sex determination testing are listed in Chapter 6: Resources. Endoscopy can also be used for sex determination if the bird requires such evaluation for poor breeding success or other health-related reasons. The following phenotypic traits can be useful in determining sex:

- *Galliformes*: Adult male birds have spurs, different plumage, and are larger than the females.

- *Columbiformes*: The males of some species will be larger. Observations of mixed sex groups may also indicate some behavioral differences of the males. A cloacal exam done under anesthesia with a speculum can help differentiate the sexes. The females have one opening on the left for the entrance of the oviduct into the cloaca, while the males have bilateral papillae associated with the ductus deferens.

- *Psittaciformes*: Most psittacines are monomorphic. The Eclectus parrot is the most prominent contrary example, with the body of the male green while the female is red. Wild type adult female cockatiels have bars on underside of the wing and tail feathers typical of immature cockatiels, while males lose the bars when they mature. Wild type male budgerigars have a blue tinted cere while the females have a flesh-colored cere. Many cockatiel and budgerigar color morphs, however, may not adhere to these rules. Some species may have very subtle differences in eye color, such as cockatoos, or feather color, such as African Greys, but these should be considered unreliable in determining sex. Cloacal exam can also be used to evaluate sex as described for columbiformes.

- *Passeriformes*: Although many passerine species are sexually dimorphic in plumage coloration, the species often used in research often do not have this characteristic (e.g., zebra finches). Males can be identified by recognizing courting behavior and vocalizations. Males of some species in breeding season will develop a cloacal promontory, a swelling of the seminal glomerulus.

- *Anseriformes*: Many duck species have marked differences in plumage between the sexes, although some differences are more subtle such as in the Pekin duck where the male has more curled tail feather but is otherwise the same color as females. Geese are difficult to differentiate based on plumage alone. If the sex is not immediately obvious, it can often be determined by observations of behavior and vocalizations. The males also have a prominent phallus which can be observed by gently pressing the cloaca.

Head and Neck

- *Ears*: The ears can be visualized by lifting forward the feathers on the side of the head, caudal to the eyes. A small otoscope can be used to evaluate the ears if needed, but ear infections are rare in adult birds.

- *Eyes*: The conjunctiva of birds is not unique and the same principles apply in its evaluation as with other species. Conjunctivitis is one of the more common ophthalmic lesions observed and most commonly results from a bacterial infection. Perioribital swelling can occur with sinusitis. Older birds can develop cataracts and other ocular lesions characteristic of aging. Ectropion can be caused by eyelid paralysis, and is common in some species such as cockatiels. Birds will have a menace and palpebral reflex, but pupillary light reflexes are not as reliable as in other species because birds can intentionally control pupil constriction and dilations due to the presence of skeletal muscle.

- *Cere/nares*: The cere should be smooth and the nares clean, symmetrically circular and with the operculum easily visible. The nares are prominent in most species, but will be underneath a superficial layer of feathers in some species such as cockatoos. A hyperkeratotic cere is abnormal and indicative of mites (*Knemidocoptes*) or malnutrition. Discharge matted to the nearby feathers is indicative of an upper respiratory infection. The shape of the nares may become dystrophic with chronic respiratory infections or malnutrition. The nares can be partially or unilaterally blocked with debris such as feather dander. Some debris may occur in birds that do not have sufficient water to bath themselves, but an underlying infection or severe malnutrition must be suspected with excess debris.

- *Beak*: The beak should be smooth and symmetrical. A small amount of flaking can occur without evidence of an underlying disease, but excessive flaking and overgrowth can be a sign of malnutrition or liver disease. *Knemidocoptes* mite infections can extend from the cere down over the beak and give a raised, rough appearance.

- *Oral cavity*: The choanal papilla size and shape vary greatly between species. Familiarity with the normal shape can be useful as blunting can be a result of malnutrition, particularly hypovitaminosis A. Abscesses associated with the choanal slit can occur secondary to upper respiratory tract infections or malnutrition. White plaques in the mouth are characteristic of *Trichomonas*, *Candida*, bacterial infections, and less commonly a poxvirus.

- *Trachea*: The trachea is not always evaluated, but the thin avian skin allows transillumination to observe for mites if suspected.

- *Crop*: The crop can be palpated at the base of the right side of the neck, but palpation should be done cautiously or avoided if it contains food.

Skin and Feathers

- *Hydration*: Skin tenting can be used to assess hydration.
- *Skin*: The skin is thin and in most areas allows visualization of the tissues beneath. The feathers are arranged in tracts and nonfeathered areas are normal, as described in Chapter 1: Important Biological Features. The skin should be

observed for masses indicative of neoplasia or an abscess. A ruptured air sac usually causes a pronounced amount of subcutaneous air that is easily noted. The skin and muscle area cranioventral to the tail is a common site of trauma, particularly for birds with impeded flight as they are unable to effectively control landing. Birds do not have palpable lymph nodes. The vent should be free of feces and urates.

- *Feathers*: Evaluate the feathers for color, conformation, and feather loss. Growing feathers are easily identified because the shaft is filled with blood. Black horizontal bars across the feathers ("stress bars") indicate that a stressor interfered with normal feather development (Jovani and Blas, 2004). A few stress bars are not a cause for concern, but excessive stress bars can indicate an underlying disease or environmental stressor. Birds are susceptible to many types of mite and lice infestations and so close observation for evidence of ectoparasites is important (see the section on "Common Diseases"). Feather loss should be noted and has many potential etiologies (see also the section on "Common Diseases" and the Chapter 2: Husbandry). Feather picking by self or cage mates can occasionally progress to severe trauma. Ulcerative or moist dermatitis seen commonly in rodents and other species is uncommon in birds with good nutritional status.

- *Uropygial gland*: For species that have a uropygial gland, it should be evaluated for masses and impactions. A small amount of sebaceous material can be expressed manually.

Musculoskeletal

- *Body condition score (BCS)*: Body condition is primarily determined by the thickness of the pectoral muscles (Figure 4.3). Some species normally have more angular pectoral muscles (between a 2 and 3 BCS) while others are more curved (between 3 and 4 BCS). Some species, particularly wild passerines, will have normal seasonal variations in weight and subcutaneous fat deposits.

- *Wings and legs*: The wings and legs should be palpated and each joint evaluated for swelling and range of motion. Swollen joints can be evidence of gout, arthritis, or an infectious disease. The sciatic nerve runs between the cranial and middle

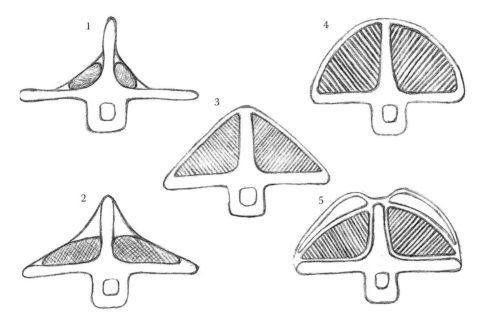

Fig. 4.3 Avian body condition scores (BCS). Cross section of the keel (vertical bone) and pectoral muscles (striped or crosshatched sections) as examples of BCS 1 through 5. (Drawing courtesy of Kristy Weed. With permission.)

pole of the kidney and along the dorsal aspect of the coelom, so renomegaly or abdominal masses can cause leg paresis or paralysis. Pododermatitis can be caused by inappropriate or insufficient perches or flooring, and is exacerbated by obesity and malnutrition (Blair, 2013; Clark et al., 2002). Other causes of lameness include trauma, arthritis, and other infectious causes of neurologic disease.

Cardiovascular and Lower Respiratory Systems

- *Heart*: The heartbeat is most audible over the middle of the keel. The fast heart rates of birds can make arrhythmias and murmurs difficult to detect, but primary cardiac disease is rare in birds. Pigeons, chickens, quails, and parrots are prone to atherosclerosis and secondary cardiac disease when fed high fat diets (Anderson et al., 2014; Beaufrère et al., 2013; Moghadasian, 2002).

- *Respiratory*: The lungs can be auscultated over the back and under the wings, but abnormal lungs sounds secondary to pathology are not as easily appreciated in birds because the lungs do not expand and contract as they do in mammals.

Coelom

- *Palpation*: Coelomic palpation in the healthy bird is unremarkable. The liver should not be palpable and individual structures are not easily identified. A distended coelom will be caused by either fluid or tissue and this must be discerned. Underlying structures may be seen through the skin, especially by wetting with alcohol if needed. Breeding females are at risk for egg yolk peritonitis when the ovum is deposited into the coelom instead of the reproductive tract.

general diagnostic tests

Fecal Exam

Fecal Gram stains are very useful in assessing birds for intestinal overgrowth of Gram-negative bacteria, clostridial infections, *Candida*, and avian gastric yeast. Most species exhibit predominantly (~80%) Gram-positive rods in normal feces. Galliformes and carnivorous birds have a higher percentage of Gram-negative bacteria. *Candida* can be present normally depending on the diet, but the presence of budding yeast suggests an active infection is present. Fecal float, direct examination, and cultures have value and use techniques similar to other species.

Urinalysis

A urinalysis is done less frequently in birds than in mammals because uncontaminated samples are difficult to obtain, but it can provide useful information. Urine can be collected by aspirating the liquid portion of excrement when it is voided on to a nonpermeable surface. It should be clear and colorless in most species. The specific gravity should be measured and the presence or absence of protein, glucose, ketones, and blood determined using commercially available dipsticks. Urine sediment should be examined for blood cells, casts, and bacteria. Some bacterial contamination is unavoidable and expected due to the comingling of feces and urine when voided.

Clinical Chemistry

Profiling avian serum chemistry values should comprise part of any patient evaluation. Lithium heparin tubes are considered suitable for blood collection and storage in most instances (Lumeij, 1987). As is always the case, samples must be handled properly and processed promptly to ensure accurate results. To that end, consultation with your diagnostic lab about proper procedure is the wisest course. Your testing lab is also the best source for normal values, although there are published values for some species which can be helpful in interpreting results (Halsema et al., 1988; Harr, 2002; Lumeij and Overduin, 1990).

The number of indices evaluated will depend largely on the presentation of the patient and judgment of the clinician. A thorough discussion of all tests is beyond the scope of this work and the focus here will be on routine testing and characteristics that separate the avian patient chemistry from other species more commonly managed. The reader is referred to chapters in major texts such as *Avian Medicine: Principles and Application* by Ritchie, Harrison, and Harrison (Hockleithner, 1994) and reviews in the published literature such as those by Harr (2002) for a thorough discussion of this subject matter. The following are tests for consideration to evaluate major organs and organ systems:

- Liver
  - Bile acids become elevated almost exclusively due to liver dysfunction and hence should always be measured in birds as part of a clinical chemistry. Some species without a gall bladder do exhibit significant elevations after meals and so fasting samples are most reliable (Lumeij and Remple, 1992).
  - Alanine aminotransferase (ALT), alkaline phosphatase (AP), aspartate aminotransferase (AST), gamma glutamyl transferase (GGT), glutamate dehydrogenase (GLDH), and lactate dehydrogenase (LDH): All are considered measures of liver health and function, but can be difficult to interpret in birds as several medical conditions can result in deviations from normal. AP levels are elevated in juvenile birds and LDH exhibits seasonal fluctuations, but both will be high secondary to liver disease. GLDH is fairly specific for the liver, and elevated levels strongly suggest hepatocellular damage making this liver enzyme the most informative in birds (Lumeij and Westerhof, 1987).

- Bilirubin increase is indicative of hepatic disease in species that form this pigment. However, chickens are one species that forms biliverdin as the major bile pigment, and hence, this test has no value in that species.
- Kidneys and Urinary Tract
  - Uric acid is most commonly analyzed to assess for renal disease in birds with hyperuricemia suggesting loss of approximately 75% of renal function. High circulating levels can lead to urate deposition in tissues, the condition known as gout.
  - Creatinine level will become elevated with severe renal disease, but it is not as reliably diagnostic as it is in mammals due to differences in metabolism.
  - Urea is not a reliable indicator of renal disease in most avian species, but does rise sharply in response to dehydration (Lumeij and Remple, 1991).
- Pancreas
  - Amylase elevation to three times normal and above suggestive of pancreatitis.
  - Lipase typically changes in conjunction with amylase.
- Nonspecific
  - Total protein provides a good indicator of overall health. Decreases can be the result of myriad causes including diminished production in the liver, malnutrition, malabsorption from the gut, or loss from the gut or kidneys to name a few. Hyperproteinemia is relatively uncommon.
  - Electrolyte (sodium, potassium, and chloride) level changes occur with many conditions ranging from dehydration and diarrhea to renal and adrenal disease. Potassium levels are susceptible to dramatic artifactual change if blood cells are not immediately separated from plasma or serum (Harr, 2002).
  - Glucose is metabolized in birds much as it is in mammals, and elevations and decreases have similar causes and results. For example, an increase will be seen postprandially or in association with diabetes mellitus, and a decrease will result from anorexia and can lead to seizure activity.
  - Calcium and phosphorous level deviations alone indicate no single disease process. Total calcium levels rise with

egg laying in females and secondary to excessive dietary vitamin D. Elevated phosphorous can indicate renal disease or parathyroid dysfunction.

Hematology

Hemogram evaluation can play an important role in evaluating the general health of an avian patient or in diagnosing the cause and progression of clinical disease. As noted earlier, lithium heparin tubes are typically preferred for blood collection. Normal avian leukocytes are generally comparable in appearance to those of mammalian species, while erythrocytes are oval or elliptical and contain a nucleus. Immature forms such as rubricytes can occasionally be found in circulation. Thrombocytes are similar to erythrocytes in appearance but are smaller with clear cytoplasm that occasionally contains red granules. Heterophils, eosinophils, and basophils comprise the granulocytic leukocytes of birds. Heterophils are comparable to neutrophils in function and appearance but have eosinophilic granules when stained with Romanowsky stains (Figure 4.4). The size and preponderance of the pink granules varies between avian species but are usually much smaller than in eosinophils. Eosinophils contain a nucleus with fewer lobes than heterophils. Basophils are slightly smaller than heterophils with a rounded nucleus and basophilic (i.e., blue) granules in the cytoplasm. Lymphocytes and monocytes are the mononuclear leukocytes found in the peripheral blood of avian species and are similar to each other in morphology, but mature monocytes are larger. Lymphocytes in some species are the most numerous circulating leukocyte (Harper and Lowe, 1998).

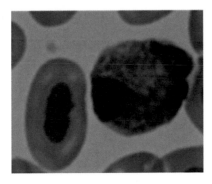

Fig. 4.4 Romanowsky stain of a blood smear from a Bengalese finch, showing a heterophil and nucleated red blood cell.

The characterization of shifts from normal values is useful in efforts to diagnose the etiology of a clinical condition. The half-life of erythrocytes is 28–45 days (Campbell, 2015) and a normal packed cell volume (PCV) is between 35% and 55%. Anemia is classified as regenerative, nonregenerative, hemolytic, or hemorrhagic as with other species and the parameters used to distinguish one from the other are likewise similar. Polychromatic erythrocytes, an increase in circulating reticulocytes, and anisocytosis are all indicative of bone marrow activity and regeneration. Alterations in the leukogram, leucocytosis, or leukopenia suggest the presence of an inflammatory or infectious process with the former more commonly resulting from chronic conditions. Heterophilia often coincides with leucocytosis and occurs for the same reasons, but heteropenia is rare in birds. Alterations in eosinophil, basophil, and lymphocyte numbers are rare (Jones, 2015). Monocytosis is also rare but can be seen in the presence of chronic infectious and granulomatous bacterial infections such as tuberculosis. It is important overall to recognize that many factors such as species, egg-laying activity (Hrabcakova et al., 2014), aging (Harper and Lowe, 1998), sex of the animals, husbandry methods, circadian rhythms, and stressors can all impact the hemogram and should be acknowledged and considered when analyzing results (Cray, 2015). A protein electrophoresis can be a useful adjunct to hematology to evaluate immunoglobulins that can be elevated in infectious diseases and is available through some diagnostic labs referenced in Chapter 6: Resources.

common clinical problems and their management

Birds are generally adept at masking observable signs of clinical disease and typically are very ill when decreased activity and alertness become evident. Weight loss and decreased body condition might present before the bird shows behavioral changes for chronic conditions. This being the case, some of the conditions described herein can be challenging to detect in birds until a thorough physical exam and diagnostic tests are conducted.

Cutaneous and Feather Diseases

Broken blood feather

New feathers have an active blood supply and colloquially are called "blood feathers." Blood feathers of the wing and tail are especially prone to trauma that results in breakage. If a blood feather becomes damaged, it should be removed even if the bleeding has been controlled to prevent a recurrence. Feather removal is best done under anesthesia. Hemostats should be placed at the base of the broken feather as close to the skin as possible and the feather pulled without bending it in any direction, so that the feather is removed completely and the shaft does not break. If the feather breaks underneath the skin, the skin must be incised to allow retrieval of the remaining feather shaft or excessive hemorrhage can result.

Cutaneous and subcutaneous masses

Birds develop neoplasia similar to other species. A uniquely avian feature is the potential for squamous cell carcinoma of the uropygial gland (Beaufrère et al., 2007). Nonneoplastic masses such a xanthomas and lipomas are frequent in some species. Xanthomas appear as yellow, subcutaneous masses that are more prevalent in psittacines. Lipomas are particularly common in budgerigars. Cutaneous papillomas are benign tumors that can occur in any species, especially on nonfeathered areas. Gout can appear as white masses under the skin along with swollen joints. Abscesses have causes similar to other species, but the avian tendency to form caseated pus must be considered when managing them.

Cysts

Cutaneous and follicular cysts are most common in small birds, often the result of trauma, genetic predisposition, or infectious agents. Cysts can become easily traumatized and surgical removal should be considered. Excision is typically curative.

Ectoparasites

- *Knemidocoptes* spp.: Budgerigars are most commonly infected, but other psittacines, passerines, chickens, and occasionally other avian species are susceptible depending on the mite species (Mete et al., 2014; Morishita et al., 2005; Pence et al., 1999; Toparlak et al., 1999). This burrowing mite causes a characteristic proliferative lesion that is typically referred to as "scaly leg" or "scaly face/beak." Lesions are most common on the cere or feet, but can be located at other areas with exposed skin. The mites are easily collected with a skin scrape of the lesion, which can be placed on a slide in mineral oil and observed microscopically for identification. Treatment with ivermectin or other avermectin antiparastics is effective.

Topical treatment can be used for focal lesions and systemic for more generalized disease.

- *Dermanyssus*: *Dermanyssus gallinae* can affect all birds, although it is commonly referred to as the "red poultry mite" or "chicken mite" with poultry most commonly infected. The mite assumes a red color once filled with blood. Clinical signs can include anemia and pruritus. The mite can be found in the environment and be zoonotic. The mites are observable with low magnification. Treat with ivermectin or pyrethrins, keeping in mind that the environment will likely also be contaminated and require treatment (Pritchard et al., 2015).

- *Feather mites*: Hundreds of feather mite species can infect birds. Feather mites can affect many species of birds and are one of the primary infectious concerns when importing birds for research (Siddalls et al., 2015). Each mite species has a tropism for a specific part of the feather, including quill mites that live inside the feather shaft. Most feather mites cause asymptomatic infections, especially in healthy adult birds. Clinical signs may appear sporadically in a colony or in young animals, and include feather abnormalities, feather loss, or dermatitis (Figures 4.2 and 4.5). Treat with pyrethrins or ivermectin, considering that efficacy may vary between mite species.

- *Lice*: Lice can affect many species of birds. Lice are typically more common in wild birds, but have been found in birds

Fig. 4.5 Bengalese finch (*Lonchura striata domestica*) with feather mites. The only clinical signs were a moth-eaten appearance to the feathers from damaged barbs.

purchased for research (Owiny and French, 2000). The animals may be asymptomatic or have mild feather abnormalities and pruritus. Like feather mites, different lice species may have an affinity for different body locations. Lice or nits can be observed grossly. Treat with pyrethrins.

- *Ornithonyssus*: *Ornithonyssus sylviarum*, the "Northern fowl mite," is another blood sucking ectoparasite of birds, primarily affecting chickens and wild birds. This mite resides on the animal for more of its life cycle than *Dermanyssus* and is usually in highest density in feathers around the vent (De La Riva et al., 2015).

Feather loss

Feather loss is a nonspecific symptom and requires an extensive workup to determine etiology. Chapter 2: Husbandry includes additional detail related to feather picking. An important distinction when evaluating a bird for feather loss is whether it results from trauma. Traumatic feather plucking can be accompanied by chewed or broken feathers or other evidence of skin trauma. However, some birds pluck their own feathers with no signs other than missing feathers. When trauma is the cause of feather loss, automutilation is more common in psittacines, while injury from cage mates is more common in chickens and passerines (Kops et al., 2014; Rubinstein and Lightfoot, 2014). Ducks that self-mutilate can also be subjected to additional trauma from conspecifics (Colton and Fraley, 2014). Behavioral feather picking is common in some of the large psittacines, such as cockatoos and African Greys, while feather picking is likely an indicator of an underlying disease in cockatiels and many smaller birds (Garner et al., 2008; Gaskins and Hungerford, 2014). Nontraumatic feather loss may be associated with dermatitis or abnormal feathers and would most likely be associated with an infectious disease. Polyomavirus should only affect young birds and is usually associated with other clinical signs. Psittacine beak and feather disease is no longer common but can cause feather lesions in adult birds. Ectoparasites can cause localized feather loss. Feather loss, whether or not due to trauma, will require the detection and elimination of the inciting cause. The diagnostic approach to feather picking is extensive and often unrewarding, but the reader can refer to the citations in this section for suggestions of how to use enrichment and other strategies to manage this condition.

Other infectious etiologies

Many species-specific avian poxviruses have been documented, including in wild-caught animals housed in a research facility (Hukkanen et al., 2003). The cutaneous form of avian poxviruses, also known as "dry pox," causes papules primarily on the unfeathered skin. This form is more common in passerines and other wild birds, but poxviruses can also affect domestic poultry (Hess et al., 2011; Zhao et al., 2014). The diptheroid form of poxviruses, also known as "wet pox," is characterized by lesions on the mucosal surface, which can cause respiratory signs. Birds can die from the septicemic form of poxvirus without obvious prior lesions, and some poxviruses cause tumors in the skin or other organs.

Digestive Disease

Infectious etiology

Bacterial

- *Chlamydophila psittaci*: *Chlamydophila* causes liver and respiratory disease. See details under Respiratory diseases.

- *Gram-negative bacteria*: Many Gram-negative bacteria can cause enteritis in birds. Some of the more common and concerning bacteria include *Escherichia coli, Campylobacter* spp., *Salmonella* spp., *Pseudomonas aeruginosa,* and *Citrobacter* spp. Infections might be limited to the digestive tract or include the liver. *Salmonella* and *Pasteurella* can affect multiple organ systems. Fowl cholera, a disease of poultry caused by *Pasteurella multocida*, causes green diarrhea secondary to liver disease. Fowl cholera might also cause chronic polysystemic signs, including respiratory and joint symptoms. A preliminary diagnosis of a Gram-negative infection can be made with a fecal Gram stain, with a fecal culture to determine specific etiology as needed. Treatments are as for other species and are ideally based on a culture and sensitivity.

- *Gram-positive infections*: *Clostridium* spp. are pathogenic for most birds and cause a severe enteritis, although Galliformes, Anseriformes, and carnivorous birds harbor small numbers in their intestinal tract normally. Further, *Clostridium perfringens* causes necrotic enteritis with associated enterotoxemia and mortality in chickens. Clostridia are easily identified on a Gram stain with the readily apparent spores. An anerobic culture can be submitted for speciation and sensitivity. Treat with appropriate antibiotics, often lincosamides, metronidazole, or penicillin. Neomycin can also be used for necrotic enteritis in chickens. *Enterococcus* are part of the normal flora, but can cause opportunistic infections.

- *Mycobacterium avium*: *Mycobacterium* is not common but important because of the zoonotic potential. This bacterium primarily affects not only the digestive tract, but may also cause granulomas and infections of other organ systems. Animals may present with chronic weight loss. Diagnosis is made with the detection of acid-fast bacteria in fecal samples, or cytology, and histopathology of the digestive tract. Euthanasia is recommended because of the zoonotic potential.

Viral

Viral infections causing digestive disease are rare in birds, but adenoviruses and herpes viruses occasionally affect the digestive tract and liver. Adenoviruses can also cause pneumonia and central nervous system (CNS) signs. Diagnosis relies on histopathology of affected tissues or PCR tests, if available (see the diagnostic laboratories listed in Chapter 6: Resources). Proventricular dilatation disease causes a wasting syndrome in psittacines and is thought to be caused by a bornavirus (Hoppes et al., 2013). Hemorrhagic enteritis is viral disease of turkeys caused by an adenovirus, but chickens are susceptible to experimental infection (Silim et al., 1978).

Parasites

Numerous endoparasites can cause digestive disease and related symptoms including weight loss, diarrhea, and anorexia. Birds with access to the ground like those housed outdoors are more likely to become infected. Infections also tend to present after shipping. A fecal float is used as the primary diagnostic test, and any additional tests are indicated hereunder.

- *Coccidia*: *Eimeria* and *Isospora* spp. have been found in many species of birds and are usually easily diagnosed on a fecal float (Figure 4.6). Clinical signs can include melena in addition to weight loss and diarrhea. In the authors' experience, amprolium is an effective agent for some coccidial species, and metronidazole and sulfa drugs can also be used. *Atoxoplasma*

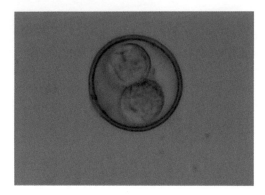

Fig. 4.6 Oocyst of unidentified coccidia from a fecal float of a Bengalese finch (*Lonchura striata domestica*) from a colony with increased mortality.

is a common coccidian parasite of passerines that is difficult to treat. Clinical signs are more likely in young birds and can range from asymptomatic to mortality. Adult birds may also develop clinical signs, especially if stressed or immuno-suppressed. The parasite also infects both the digestive tract and the liver. Although oocysts might be found on a fecal float, they are shed intermittently and therefore false negative results are common. Postmortem diagnosis can be made by noting intracytoplasmic inclusion bodies in the liver or digestive tract. No effective treatments are currently available in the United States. The agent can be very difficult to eliminate from a colony once established.

- *Protozoa*: *Giardia*, *Spironucleus* (*Hexamita*), *Cochlosoma*, and *Trichomonas* all infect the digestive tract of birds. Signs can range from asymptomatic to diarrhea, feather picking, and mortality. *Trichomonas* is unique among these parasites in causing white plaques within the oral cavity. Diagnosis can be difficult, but the chance of detecting the organism is increased with both a fecal direct and a zinc sulfate fecal float. Treat with metronidazole. *Histomonas meleagridis* is the causative agent of blackhead disease in poultry, which can cause a fatal typhlitis and hepatitis.

- *Nematodes*: Ascarids are one of the more common parasites of outdoor birds. *Heterakis gallinarum* is an ascarid cecal worm of gallinaceous birds which can cause mild clinical signs.

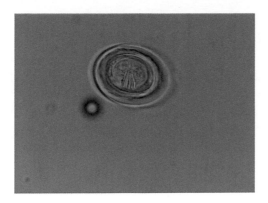

Fig. 4.7 Tapeworm egg from a fecal float of a Bengalese finch (*Lonchura striata domestica*).

> *Capillaria* infects most species and causes melena and regurgitation, in addition to diarrhea. Most nematodes will respond to pyrantel pamoate. *Capillaria* should be treated with ivermectin. Fenbendazole can be used if necessary but is generally avoided, as toxicity has been noted in some species with high doses (Gozalo et al., 2006).

- *Cestodes*: One of the more common parasites of outdoor birds. Proglottid pieces might be found associated with the cloaca or feces or eggs might be seen on a fecal float (Figure 4.7). Treat with praziquantel.

Fungal

- *Candida*: *Candida albicans* infections are typically opportunistic and secondary to other pathogenic infections, stress, malnutrition, or antibiotic treatment. Infections are more common in the crop of young birds, particularly with delayed crop emptying times. The fungus can cause white plaques in the oral cavity or be restricted to the upper gastrointestinal tract. Treat with nystatin. Azole antifungals can be used if systemic absorption is necessary.

- *Avian gastric yeast (Macrorhabdus)*: This fungus is primarily located in the proventriculus but is often seen in the fecal Gram stain of affected birds. The most common clinical sign is chronic weight loss. Small psittacines and passerines, including zebra finches, are more likely to be infected (Snyder et al., 2013). Treat with azole antifungals.

Noninfectious etiology
Hepatic lipidosis

Obese birds are prone to hepatic lipidosis. The disease may occur in conjunction with atherosclerosis. Clinical signs are those associated with liver failure and can include lethargy, green urates and urine, ascites, feather picking, and hepatic encephalopathy. An enlarged liver might be detected on physical examination. Dietary fat restriction, in addition to supportive care to address the immediate clinical concerns, is necessary. The principles of treatment and management are the same as those for other species.

Obstruction

Psittacines have a tendency to consume foreign objects that are either part of enrichment or objects inadvertently left near the cage. Any enrichment involving rope must be carefully inspected and removed once frayed. Chickens show dietary indiscretion while foraging when small loose objects are available on the ground. The excessive consumption of grit will result in impactions.

Respiratory Diseases/Conjunctivitis

General management of respiratory disease

Treatment for acute respiratory disease is similar to that in other species, with a focus on providing supplemental oxygen and minimizing stress to the animal. Unique to birds, an air sac tube can be used instead of endotracheal intubation if needed (see below under "Intubation"). For upper respiratory disease, a nasal flush is useful both to clear the nares and obtain a sample for diagnostic tests (see Chapter 5: Experimental Methodology). Analgesics, bronchodilators, and additional supportive care in the form of nutrition and warmth can also help improve clinical outcomes.

Infectious etiology
Viral

Viral causes of respiratory disease are likely uncommon in laboratory birds. Viral etiologies do not typically localize to the respiratory tract and often are associated with polysystemic signs. Adenovirus, such as fowl adenovirus and quail bronchitis, can cause pneumonia in addition to infecting the gastrointestinal tract, liver, and brain. Avian influenza is highly unlikely in laboratory birds but can cause respiratory signs, in addition to diarrhea and CNS signs, and could become a differential diagnosis in wild-caught animals and chickens depending on the source. Avian influenza type A viruses can cause respiratory disease along with lethargy, hemorrhage, cyanosis, and occasionally neurologic signs. The severity of the disease depends on the virus, with low pathogenic viruses causing less severe disease than highly pathogenic viruses, and the species affected. Diagnosis requires isolation of the virus from infected tissues, such as the respiratory tract and liver. Herpes viruses, such as infectious laryngotracheitis in galliformes, can cause upper respiratory disease and conjunctivitis or even death. Paramyxoviruses, which include the agent causing Newcastle disease, cause CNS signs dependent on the viral strain in addition to respiratory disease. Poxviruses can cause lesions in the respiratory tract, possibly in conjunction with cutaneous or systemic signs (see "Cutaneous diseases"). Infectious bronchitis is a highly contagious coronavirus affecting chickens.

Bacterial

A variety of bacteria are associated with upper or lower respiratory disease in birds, including *Klebsiella*, *Pasteurella*, and *Pseudomonas*. *Haemophilus paragallinarum* is the cause of infectious coryza in chickens and is associated with upper and lower respiratory disease as well as air sac disease. *Mycoplasma* causes conjunctivitis and sinusitis in small psittacines, passerines, columbiformes and galliformes.

Chlamydophila psittaci is an obligate intracellular bacterium that causes respiratory and liver disease. Clinical signs will be associated with these organ systems and include conjunctivitis, nasal discharge, dyspnea, green urates and urine, weight loss, and diarrhea. The upper respiratory tract symptoms can be severe. Rarely the disease will include CNS signs, with or without involvement of the liver and respiratory tract. Some bird species, such as cockatiels, budgerigars, and small passerines, can be inapparent carriers of the disease. Serology and PCR tests are available for diagnosis. A single swab of the choana and cloaca, along with blood is recommended for PCR to maximize the chance of finding the agent. Serology is useful as a screening tool to determine if new arrivals have been exposed to the bacteria. Paired serology might be necessary, as low titers might either be truly insignificant (designated as tentative or negative by the laboratory), or, conversely, might be present in recently exposed animals that have not yet responded robustly. Treatment is not recommended because of the zoonotic potential.

Fungal

Aspergillosis is opportunistic and more likely in stressed or immunocompromised animals, particularly if housed outdoors or in poorly cleaned environments. The fungus can infect any part of the respiratory tract, and will occasionally disseminate to the liver, gastrointestinal tract, and other organs. Granulomas within the trachea can cause severe dyspnea if present. A serology test is available to aid in diagnosis. Postmortem diagnosis can be made by demonstration of the organism within white plaques. Treatment includes prolonged administration of antifungals, and may not be realistic in a lab environment.

Parasites

Air sac mites are more common in small birds. The mites live in the air sacs, trachea, and bronchi. Birds commonly show respiratory signs, but severity will vary. They may be visualized in the trachea with transillumination on the physical exam. *Sternostomoa tracheacolum* is a common respiratory tract mite of canaries and finches. Treat with ivermectin.

Noninfectious etiology

Respiratory distress can occur from any space occupying mass impinging on the airway, in addition to primary respiratory disease. Tracheal obstructions can occur from inhaled foreign bodies, goiter, or granulomas. A tracheal membrane and/or stenosis can occur subsequent to intubation. The bird will likely be present with audible inspiratory wheezes and dyspnea but no evidence of other respiratory disease. An air sac tube should be placed to stabilize the animal. Treatment includes a tracheal resection and anastomosis. Cardiovascular disease can present as respiratory distress, such as from primary cardiac disease, atherosclerosis, or avocado toxicity (Burger et al, 1994; Costa et al., 2013; Ryan, 1992). Inhaled toxins can also cause respiratory distress. Polytetrafluoroethylenes (PTFE) can cause pulmonary edema and liver failure. Although this toxicant is classically associated with overheating nonstick cookware, laboratory birds have also been affected by bulbs coated with this agent (Shuster et al., 2012). Neoplasia of any consequence to the respiratory system must be considered as in other species (Azmanis et al., 2013).

Reproductive Diseases

Egg binding

Retained eggs are more common in birds with a history of egg laying, particularly an excessive number of unfertilized eggs and/or if the animal is on a low calcium or high phosphorus diet. The egg should be easily palpable as a hard mass in the coelom, and the bird will likely either be on the bottom of the cage or have a wide legged stance on the perch. Egg bound birds are typically lethargic, fluffed, and dehydrated. Radiographs are highly recommended prior to treatment, in the event multiple eggs are present but difficult to detect on palpation. The bird should be stabilized in a warm and quiet environment, with subcutaneous fluids, calcium, nutritional support and analgesics as deemed appropriate. Manual removal under sedation, implosion via ovocentesis, or even surgical removal, may be required.

Neurologic Diseases

Infectious etiology

Neurologic diseases with an infectious etiology are unlikely in a laboratory environment and are of greater concern in commercial poultry farms, but the reader should be aware of some classic pathologic agents affecting this system. Newcastle disease virus, an avian paramyxovirus, can cause CNS signs in addition to respiratory symptoms and diarrhea in many species of birds. Avian influenza type A viruses can be associated with neurologic signs in addition to the more common respiratory signs and can also infect both wild and domestic birds. Marek's disease is caused by an alphaherpesvirus that induces lymphoid tumor formation. The presenting symptoms coincide with tumor location, which may be the CNS, paralysis, ocular, visceral, or cutaneous signs. West Nile virus is another disease of wildlife with signs ranging from asymptomatic to severe CNS disease. Although these viruses will be extremely rare findings in laboratory birds, most are on the list of United States Department of Agriculture (USDA) reportable diseases if they are diagnosed. Proventricular dilatation disease can cause CNS signs or sudden death, in addition to the classic weight loss and digestive disease. Botulism caused by toxin from *Clostridium botulinum* should only be a consideration in birds housed outdoors, especially waterfowl, although other birds can be affected (Anniballi et al., 2013; Popp et al., 2012).

Noninfectious etiology

The rule-outs and workup of seizures and other neurologic signs in birds are similar to that of other species. The environment should be evaluated for potential sources of toxins, such as heavy metals (lead and zinc) in the cage or furnishing and aflatoxins in the food. Liver failure and hypocalcemia are a potential cause, particularly if birds are on a poor diet. Neoplasia, hypoglycemia, and hyperthermia can also present with CNS signs.

Miscellaneous Diseases

Malocclusion

Malocclusion can occur in the form of abnormal lateral deviations ("scissor beak") or mandibular prognathism or brachygnathism. These conditions can be congenital or arise from neonatal trauma. A healthy adult bird should not require beak trims to function normally, and overgrown beaks in adults without a history of malocclusion suggest malnutrition. Brief beak trims can be done in awake animals with a rotary tool (e.g., Dremel®). Extensive trimming is best done under anesthesia, especially for very small birds such as passerines which can succumb to the stress associated with a beak trim.

Splay leg

Splay leg will occur in neonatal birds due to genetic predisposition or poor housing conditions. The deformity can be corrected by hobbling the legs together, but not all animals respond to this treatment. The bandage should be changed weekly to allow inspection of the skin and prevent abnormal positioning as the animal grows.

Traumatic injury

In the laboratory setting, most injuries arise from inadequate housing, startle responses, or conspecifics resulting in a range of injuries from simple bruises and lacerations to fractured bones. If a fractured leg or wing is diagnosed, the caging should be inspected for areas in which the limb could have been lodged or for obstructions to flight. The principles of fracture repair are similar to that of other species. If surgical repair is not realistic, a fracture distal to the elbow can be stabilized with a figure 8 bandage (Figure 4.8). The figure 8 bandage can include a wrap around the body to stabilize the shoulder joint for a humeral fracture. Species that use their beak for movement around a cage (e.g., psittacines) can function with one leg if an amputation is

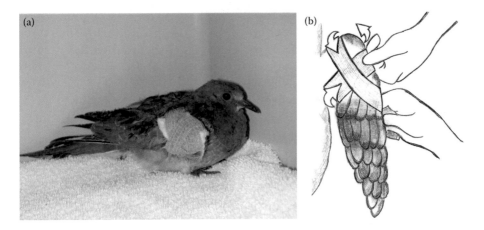

Fig. 4.8 (a) A figure eight bandage on a dove. (b) A figure eight bandage is placed by wrapping the top part of the "8" around the wrist and the bottom part around the elbow and flight feathers. Although this bandage will only stabilize fractures distal to the elbow, the bandage can be secured by wrapping around the body for stabilization, without including the opposite wing. (Drawing courtesy of Kristy Weed. With permission.)

necessary, but most species struggle with only one leg. The prognosis for a wing amputation is good if the bird is housed such that it does not need to fly. Enough muscle must be left over the amputation site to avoid pressure and damage to the bone.

Lacerations and bruising over the keel and tail base can occur in birds who try to fly but land awkwardly, due to clipped or damaged wings. Toe devitalization and necrosis can occur secondary to string restricting blood flow. If necessary, toe amputations are easily accomplished in birds. Trauma from cage mates can be severe in some species, such as chickens and larger psittacines. The compatibility of the animals and environment should be evaluated for contributing factors. Analgesic options to consider for traumatic injuries are discussed in the subsequent section.

anesthesia and analgesia agents

The safety and efficacy for anesthetic and analgesic agents can vary widely between species, and therefore drug doses are beyond the scope of this chapter. The formularies listed in Chapter 6: Resources provide more information. General anesthesia should be considered for any procedure which is expected to cause excessive stress or pain to an animal. Minor procedures such as wing trims and nail trims are typically done in awake animals, but the stress associated with restraint may make general anesthesia desirable over local anesthesia for even minor invasive procedures, such as a feather biopsy. Inhalant agents are the preferred and most common anesthetic agents for birds, even for procedures in which the air sacs are exposed. While injectable agents may be incorporated into the anesthetic protocol, reliable and safe regimens are not as well established as in many mammalian species. A few injectable protocols have been developed for imaging and other procedures where isoflurane can be difficult to use, although supplemental oxygen is recommended when possible.

Analgesics and Anesthetics

Opioids

Kappa receptor agonists are the preferred opioids in birds. In mammals, kappa receptors are responsible for mediating pain at the level of the spinal cord while mu receptors mediate supraspinal analgesia. In contrast, pigeons have a greater distribution of kappa receptor in the forebrain and midbrain than mu receptors (Reiner et al., 1989). Studies of mu receptor agonists in birds have shown limited and variable efficacy, while kappa agonists are considered more effective (Hoppes et al., 2003; Hughes, 1990; Paul-Murphy, et al., 1999, 2009; Sladky et al., 2006). Butorphanol is a commonly used opioid in birds, although the frequent dosing required makes it more useful in the immediate perioperative period than for longer term pain management (Curro et al., 1994; Reim and Middleton, 1995; Riggs et al., 2008; Singh et al., 2011).

NSAIDs

Meloxicam is the most commonly used nonsteroidal anti-inflammatory drug (NSAID) in birds and should be used with the same precautions and considerations as in other species. Birds often readily accept the oral formulation once they are adapted to taking medications from a syringe. Other NSAIDs can be useful for very brief pain management and are likely sufficient for suspected mild pain caused by minor procedures.

Inhalants

MAC is used for dosing inhalant anesthesia in birds, although it refers to "minimal anesthetic concentration" since birds do not have an alveolar lung. The MAC is the minimal anesthetic concentration required to prevent movement in response to a noxious stimulus in 50% of the birds.

Isoflurane is the most commonly used inhalant anesthetic due to cost and availability of compatible equipment. While isoflurane causes less respiratory depression than halothane in birds, the respiratory depression with isoflurane is more pronounced than in mammals (Ludders et al., 1990). Sevoflurane is also used for birds and may be preferable if available because the less noxious odor might decrease the likelihood of breath holding.

Benzodiazepenes

Benzodiazepenes are gamma-aminobutyric acid (GABA) agonists used for sedation, premedication or as part of an injectable anesthesia regimen. Diazepam and midazolam have both been used in birds, but midazolam is preferable because it is better absorbed via the intramuscular route and accessing veins in birds can be challenging. Midazolam can be successfully used for mild to heavy sedation with minimal cardiovascular side effects and is typically given via the intramuscular, and occasionally intranasal, route (Day and Rogem, 1996; Mans et al., 2013; Valverde et al., 1990). The level of sedation and recovery time can be variable between species, so the clinician should use a conservative dose until familiar with use of the drug, particularly in sick animals. In the authors' experience, midazolam can be combined with ketamine for surgical anesthesia in small passerines, although the combination does not uniformly provide reliable anesthesia in many species and even in passerines has the risk of respiratory depression (Maiti et al., 2006).

Dissociatives

Ketamine is the most common N-methyl-D-aspartate (NMDA) antagonist used in birds. Ketamine alone will not induce a surgical plane of anesthesia, and generally should not be used alone. Ketamine combinations can be used successfully for deep sedation or anesthesia, but personnel must be familiar with the protocols in the particular species they are using as these combinations often have a narrow margin of safety. Ketamine has been combined with alpha2 agonists for imaging and surgical procedures in passerines and xylazine for surgical procedures in chickens, although they still must be used with caution because of the potential for serious side effects (Boumans et al., 2007; Maiti et al., 2006; Muresan et al., 2008; Van Meir et al., 2005). Ketamine has been combined with benzodiazepenes for surgical anesthesia in small passerines as well (see "Benzodiazepenes").

Alpha2 receptor agonists

Xylazine and dexmedetomidine are most commonly considered, but these are not appropriate as sole agents for sedation or anesthesia (Sandmeier, 2000). They have been used effectively in some species with some caveats (see "Dissociatives"). Alone the agents have highly variable responses and potentially severe side effects (Uzun et al., 2006).

Propofol

Propofol is a GABA agonist suitable for a rapid induction, but it requires intravenous administration making it an infrequent choice. The recovery time can be relatively fast in some species (Kembro et al., 2012). Potential complications include severe respiratory depression, prolonged recovery, a narrow margin of safety, and CNS excitement in some species, with effects partially dependent on method of administration and monitoring capabilities (Fitzgerald and Cooper, 1990; Hawkins et al., 2003; Machin and Caulkett, 1998, 2000; Müller et al., 2011).

Barbiturates

Barbiturates, including phenobarbital and pentobarbital, are GABA agonists which historically have been used for experimental procedures in birds. These agents have a narrow margin of safety and are no longer recommended anesthetic agents. The experimental studies which typically involve birds can be conducted with safer anesthetic agents that are now available.

Anesthetic Procedures and Considerations

Fasting

Birds must be fasted prior to anesthesia because they can maintain food within the crop and easily regurgitate creating a risk for aspiration of crop contents. The goal with fasting is to allow the crop to empty, and smaller birds should be fasted approximately 3 h while larger birds can be fasted up to 8 h.

Premedication and induction

The stress associated with handling and injections should be considered when determining whether or not to administer a premedication prior to inhalant anesthesia, especially in small passerines. The benefit of premedication may outweigh the stress associated with injections for larger birds. Clinicians vary considerably in opinions regarding the stress associated with isoflurane induction, and little research exists that evaluates this particular aspect of isoflurane anesthesia. Butorphanol is the most commonly used premedication because it also provides analgesia and can be combined with midazolam for additional sedation. This combination at commonly used doses will only provide mild sedation. Atropine is not a necessary part of premedication and has the potential to thicken tracheal secretions that could obstruct small endotracheal tubes.

Inhalant agents are predominantly used for induction. Larger birds should be hand restrained and mask induced with the inhalant agent. Small birds can be induced in an induction box attached to a vaporizer, as designed for small rodents. The animal must be carefully watched for regurgitation and movements that could result in harm to the bird.

Anesthesia systems

A nonrebreathing circuit is recommended for most birds because of their small size. Although nonrebreathing circuits require higher oxygen flow rates than rebreathing systems, the small size of most patients minimizes the cost impact.

Intubation

Tracheal intubation is easily accomplished in larger birds. Intubation of birds weighing less than 100 g is controversial since endotracheal tubes are more likely to become obstructed with mucus. Birds have complete tracheal rings, and therefore noncuffed tubes should be used to avoid excessive pressure on the trachea. Cuffed tubes can be used without inflating the cuff, but the uninflated cuff could increase tracheal irritation. If intubation is desired for birds less than 100 g, small endotracheal tubes (see Chapter 6: Resources) or IV catheters can be used. The tube will likely be much longer than necessary, and should be measured so that it is placed cranial to the thoracic inlet.

The larynx is easily visualized at the base of the tongue with the mouth open. A largyngoscope can be used to help illuminate the oral cavity and visualize the larynx in birds with a deeper oral cavity. The tube should be lubricated and prepared with tape or stretch gauze. The tube can then be taped to the beak or tied around the back of the head as is done for mammals. Take care to minimize movements while the animal is attached to an anesthetic circuit, as birds

Fig. 4.9 A catheter placed as an air sac tube in a Bengalese finch (*Lonchura striata domestica*). An air sac tube can be placed in an emergency or to deliver inhalant anesthetics.

are susceptible to tracheal strictures secondary to trauma from the endotracheal tube.

Air sac intubation is an alternative to tracheal intubation and a ventilator can still be used if desired, although blood gases may be required to ensure adequate ventilation if a bird is apneic (Figure 4.9; Paré et al., 2013). Waste gases will exit the nose and mouth and scavenger systems must be positioned accordingly. The caudal thoracic or abdominal air sac should be used. The approach described below will put the catheter in one of these air sacs with the exact location dependent on the anatomy of that species and slight variations in angle or positioning. The preferred size and type of equipment will vary with the size of the bird. The technique is also described in many of the general avian medicine books cited in this section and in the lab animal literature (Brown, 2006; Nilson et al., 2005).

Equipment:

- Air sac tube: endotracheal tube, red rubber catheter, specialized avian air sac tubes, or intravenous catheters (small birds)
- Small scissors or scalpel blade
- Hemostats
- Tape
- Suture
- Optional: Tissue glue

Technique:

- Lay the animal in a lateral position and pull the top leg backward.

- Palpate the location for entry, just caudal to the last rib and ventral to the flexor cruris medialis muscle.
- Prepare the area with a surgical scrub. You should have to pick few if any feathers.
- Hemostat technique:
  - Incise the skin to create an opening just larger than the air sac tube.
  - Use the hemostats to separate the muscle and then push through the body wall. Aim slightly cranially and brace your hands such that you can stop the hemostats as soon as you no longer meet resistance, to keep from damaging internal structures.
- Catheter technique:
  - A catheter can be placed in small birds with the stylet instead of an incision and dissection with hemostats. Take care to penetrate only the body wall and prevent damage to internal organs.
- Verify tube placement using a feather or light paper to observe air movement since air flow will likely be undetectable otherwise.
- Secure the tube by placing tape on either side in a butterfly fashion and then suturing to the skin.
- Remove the catheter and oppose the skin with suture or tissue glue. The muscle layer can be closed but often is left to heal without closure. Subcutaneous emphysema might develop if the muscle is not closed, but the risk is small.

Catheters

Details on catheter placement can be found in Chapter 5: Experimental Methodology. Catheters for fluid administration can be placed in an emergency or when animals would benefit from fluid support as dictated by the nature of the procedure. More often, fluids are administered postoperatively via the subcutaneous route. Intravenous catheters are easiest to place in the jugular (with the exception of columbiformes) or medial metatarsal vein, depending on the size of the animal. The ulnar vein is difficult to catheterize because the accessible area is very short. When intravenous catheters are impractical, intraosseous catheters can be placed in the ulna or tibiotarsus. Catheters should never be placed in the humerus or femur as they are pneumatic bones.

Monitoring and supportive care

- Clear drapes allow visual monitoring of the patient.

- *Reflexes*: wing tone, toe pinch, and palpebral reflexes should be abolished when a surgical plane of anesthesia is reached.

- *Body temperature*: 105–107°F. Elevations resulting from restraint of a conscious animal is not uncommon (Greenacre and Lusby, 2004). Temperature can be monitored via probes designed for rodents or with a standard human thermometer for larger birds.

- *Heart rate*: A Doppler over the ulnar artery near the elbow and secured with tongue depressors can be used to measure the heart rate (Figure 4.10). Normal heart rates vary greatly depending on the size of the bird and can increase by over 100% with restraint. For example, a small passerine resting heart rate would be approximately 250 beats per minute, but up to 600 when restrained (Ritchie et al., 1994).

- *Electrocardiogram*: Clips can be attached to needles or wires penetrating the skin lateral to each stifle and the base of each wing.

- *Blood pressure*: Blood pressure can be determined with a cuff and a Doppler, with the cuff over the humerus and the Doppler on the ulnar artery near the medial aspect of the carpus, or the

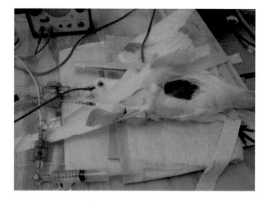

Fig. 4.10 A pigeon (*Columba livia*) prepared for coelomic endoscopy. Note the wings secured with self-adhesive bandaging tape and a Doppler probe secured over the ulnar artery with taped tongue depressors for auditory monitoring of the heart rate. An endotracheal tube attached to a ventilator, direct blood pressure, and esophageal temperature probe are also attached to the animal.

cuff over the femur with the Doppler over the metatarsal artery, although the accuracy of this technique in birds may vary in some species (Johnston et al., 2011; Lichtenberger, 2005a,b).

- *Respiratory rate*: Respiratory rate can be set if the bird is on a mechanical ventilator or observed by motion of the keel. Normal respiratory rates also vary widely with the size of the bird and can double with restraint (Ritchie et al., 1994).

- *Pulse oximetry*: Pulse oximeters are designed for mammals and therefore are not as accurate in birds, but can be useful in trending.

- *Thermal support*: Supplemental heat is critically important in birds, particularly small birds and for long procedures. The same heat sources that are used for small mammals can be used in birds.

- *Ventilators*: Ventilators are useful for long procedures as respiratory complications are more likely in birds than mammals. Respiratory rate can be set at 20–40 breaths per minute. Examples are discussed in Chapter 6: Resources.

- *Fluid support*: The fluids should be warmed because a constant infusion of fluids at room temperature will cause hypothermia. An estimate of 5–10 mg/kg/h is acceptable during surgery, with boluses as needed to compensate for blood loss.

Anesthetic complications

- *Apnea*: Birds are more prone to hypoxemia during anesthesia than mammals because they are at a higher risk for apnea and they deplete oxygen more quickly (Lichtenberger and Ko, 2007). Their lungs lack the residual volume of the mammalian lung and most anesthetics will cause more respiratory depression in birds than in mammals. Positioning in dorsal recumbency increases the risk of respiratory depression because of the effort required to raise the keel. If apnea occurs, mechanical ventilation should be instituted in the intubated patient. A tracheal or air sac tube should be placed if the bird is not already intubated. In an emergency, the anesthetic should be discontinued and any injectable agent reversed if possible (opioids and benzodiazepenes). The animal should be maintained on 100% oxygen if available until stable.

- *Stress response*: A stress response referred to as the "dive response" can occur, particularly in anseriformes. This can occur during or immediately after the mask induction of an animal, and manifests as bradycardia and apnea. If this occurs, anesthesia should be discontinued immediately and the animal intubated if necessary.

euthanasia

The most appropriate technique for euthanasia of birds is dependent on the size of the animal and the research conditions. The 2013 AVMA Guidelines for the Euthanasia of Animals considers intravenous barbiturates acceptable and inhaled anesthetics, carbon dioxide, carbon monoxide, nitrous gas, and argon acceptable with conditions. Additional physical methods are also acceptable with conditions in small birds (decapitation and cervical dislocation) and poultry (cervical dislocation).

Intravenous or intracardiac barbiturates delivered to an anesthetized animal might be a preferred method both for ease of use and impact to the animal, unless the bird can be easily restrained and the medial metatarsal vein accessed without anesthesia. The dorsal occipital venous sinus may be the easiest method of administering euthanasia solution intravenously, especially for smaller birds. Carbon dioxide chambers designed for rodents are suitable for use with small birds. Inhaled anaesthetics pose the same complications in birds as in mammals, and a secondary method of euthanasia is advised.

Decaptitation or cervical dislocation is required for some studies. Physical methods require more extensive training and can be challenging in larger poultry. A program for assuring personnel are proficient in the technique is required by the 2011 Guide for the Care and Use of Laboratory Animals. Thoracic compression has historically been commonly used for field studies, but is unacceptable in the 2013 AVMA Guidelines.

references

Anderson, J.L., Smith, S.C., and Taylor, R.L. Jr. 2014. The pigeon (Columba livia) model of spontaneous atherosclerosis. *Poultry Science.* 93(11):2691–2699.

Anniballi, F., Fiore, A., Löfström, C. et al. 2013. Management of animal botulism outbreaks: From clinical suspicion to practical countermeasures to prevent or minimize outbreaks. *Biosecurity and Bioterrorism.* 11:S191–S199.

Azmanis, P., Stenkat, J., Hübel, J., Böhme, J., Krautwald-Junghanns, M.E., and Schmidt, V. 2013. A complicated, metastatic, humeral air sac cystadenocarcinoma in a timneh African grey parrot (*Psittacus erithacus timneh*). *Journal of Avian Medicine and Surgery.* 27(1):38–43.

Beaufrère, H., Ammersbach, M., Reavill, D.R., Garner, M.M., Heatley, J.J., Wakamatsu, N., Nevarez, J.G., and Tully, T.N. 2013. Prevalence of and risk factors associated with atherosclerosis in psittacine birds. *Journal of the American Veterinary Medical Association.* 242(12):1696–1704.

Beaufrère, H., Brasseur, G., and Heimann, M. 2007. What is your diagnosis? Squamous cell carcinoma of the uropygial gland. *Journal of Avian Medicine and Surgery.* 21(4):321–324.

Blair, J. 2013. Bumblefoot: A comparison of clinical presentation and treatment of pododermatitis in rabbits, rodents, and birds. *Veterinary Clinics of North America: Exotic Animal Practice.* 16(3):715–735.

Boumans, T., Theunissen, F.E., Poirier, C., and Van Der Linden, A. 2007. Neural representation of spectral and temporal features of song in the auditory forebrain of zebra finches as revealed by functional MRI. *European Journal of Neuroscience.* 26(9):2613–2626.

Brown, C. 2006. Air sac cannula placement in birds. *Laboratory Animals.* 345(7):23–24.

Burger, W.P., Naudé, T.W., Van Rensburg, I.B., Botha, C.J., and Pienaar, A.C. 1994. Cardiomyopathy in ostriches (*Struthio camelus*) due to avocado (*Persea americana var. guatemalensis*) intoxication. *Journal of the South African Veterinary Association.* 65(3):113–118.

Campbell, T.W. 2015. Exotic animal hematology. *The Veterinary Clinics of North America: Exotic Animal Practice.* 18:xi–xii.

Clark, S., Hansen, G., McLean, P., Bond, P. Jr., Wakeman, W., Meadows, R., and Buda, S. 2002. Pododermatitis in turkeys. *Avian Diseases.* 46(4):1038–1044.

Colton, S., and Fraley, G.S. 2014. The effects of environmental enrichment devices on feather picking in commercially housed Pekin ducks. *Poultry Science.* 93(9):2143–2150.

Costa, T., Grífols, J., and Perpiñán, D. 2013. Endogenous lipid pneumonia in an African grey parrot (*Psittacus erithacus erithacus*). *Journal of Comparative Pathology.* 149(2–3):381–384.

Cray, C. 2015. Reference intervals in avian and exotic hematology. *The Veterinary Clinics of North America Exotic Animal Practice.* 18:105–116.

Curro, T.G., Brunson, D.B., and Paul-Murphy, J. 1994. Determination of the ED50 of isoflurane and evaluation of the isoflurane-sparing effect of butorphanol in cockatoos (*Cacatua* spp.). *Veterinary Surgery.* 23(5):429–433.

Day, T.K., and Roge, C.K. 1996. Evaluation of sedation in quail induced by use of midazolam and reversed by use of flumazenil. *Journal of the American Veterinary Medical Association.* 209(5):969–971.

De La Riva, D.G., Soto, D., and Mullens, B.A. 2015. Temperature governs on-host distribution of the northern fowl mite, *Ornithonyssus sylviarum* (Acari: Macronyssidae). *Journal of Parasitology.* 101(1):18–23.

Fitzgerald, G., and Cooper, J.E. 1990. Preliminary studies on the use of propofol in the domestic pigeon (*Columba livia*). *Research in Veterinary Science.* 49(3):334–338.

Garner, M.M., Clubb, S.L., Mitchell, M.A., and Brown, L. 2008. Feather-picking psittacines: Histopathology and species trends. *Veterinary Pathology.* 45(3):401–408.

Gaskins, L.A., and Hungerford, L. 2014. Nonmedical factors associated with feather picking in pet psittacine birds. *Journal of Avian Medicine and Surgery.* 28(2):109–117.

Gozalo, A.S., Schwiebert, R.S., and Lawson, G.W. 2006. Mortality associated with fenbendazole administration in pigeons (*Columba livia*). *JAALAS.* 45(6):63–66.

Greenacre, C.B., and Lusby, A.L. 2004. Physiologic responses of Amazon parrots (*Amazona* species) to manual restraint. *JAMS.* 18(1):19–22.

Halsema, W.B., Alberts, H., de Bruijne, J.J., and Lumeij, J.T. 1988. Collection and analysis of urine from racing pigeons (*Columba livia domestica*). *Avian Pathology: Journal of the WVPA.* 17:221–225.

Harper, E.J., and Lowe, B. 1998. Hematology values in a colony of budgerigars (*Melopsittacus undulatus*) and changes associated with aging. *The Journal of Nutrition.* 128:2639S–2640S.

Harr, K.E. 2002. Clinical chemistry of companion avian species: A review. *Veterinary Clinical Pathology/American Society for Veterinary Clinical Pathology.* 31:140–151.

Hawkins, M.G., Wright, B.D., Pascoe, P.J., Kass, P.H., Maxwell, L.K., and Tell, L.A. 2003. Pharmacokinetics and anesthetic and cardiopulmonary effects of propofol in red-tailed hawks (*Buteo jamaicensis*) and great horned owls (*Bubo virginianus*). *American Journal of Veterinary Research.* 64(6):677–683.

Hess, C., Maegdefrau-Pollan, B., Bilic, I., Liebhart, D., Richter, S., Mitsch, P., and Hess, M. 2011. Outbreak of cutaneous form of poxvirus on a commercial turkey farm caused by the species fowlpox. *Avian Diseases.* 55(4):714–718.

Hockleithner, M. 1994. Biochemistries. In *Avian Medicine: Principles and Application*, B.W. Ritchie, G.J. Harrison, and L.R. Harrison, eds. (Lake Worth, Florida: Wingers Publishing), pp. 223–245.

Hoppes, S.M., Tizard, I., and Shivaprasad, H.L. 2013. Avian bornavirus and proventricular dilatation disease: Diagnostics, pathology, prevalence, and control. *Veterinary Clinics of North America: Exotic Animal Practice.* 16(2):339–355.

Hoppes, S., Flammer, K., Hoersch, K., Papich, M., and Paul-Murphy, J. 2003. Disposition and analgesic effects of fentanyl in white cockatoos (*Cacatua alba*). *Journal of Avian Medicine and Surgery.* 17(3):124–30.

Hrabcakova, P., Voslarova, E., Bedanova, I., Pistekova, V., Chloupck, J., and Vecerek V. 2014. Haematological and biochemical parameters during the laying period in common pheasant hens housed in enhanced cages. *The Scientific World Journal.* 2014, 364602.

Hughes, R.A. 1990. Strain-dependent morphine-induced analgesic and hyperalgesic effects on thermal nociception in domestic fowl (*Gallus gallus*). *Behavioral Neuroscience.* 104(4):619–624.

Hukkanen, R.R., Richardson, M., Wingfield, J.C., Treuting, P., and Brabb, T. 2003. *Avipox* sp. in a colony of gray-crowned rosy finches (*Leucosticte tephrocotis*). *Comparative Medicine.* 53(5):548–552.

Johnston, M.S., Davidowski, L.A., Rao, S., and Hill, A.E. 2011. Precision of repeated, Doppler-derived indirect blood pressure measurements in conscious psittacine birds. *Journal of Avian Medicine and Surgery.* 25(2):83–90.

Jones, M.P. 2015. Avian hematology. *The Veterinary Clinics of North America Exotic Animal Practice.* 18:51–61.

Jovani, R., and Blas, J. 2004. Adaptive allocation of stress-induced deformities on bird feathers. *Journal of Evolutionary Biology.* 17(2):294–301.

Kembro, J.M., Guzman, D.A., Perillo, M.A., and Marin, R.H. 2012. Temporal pattern of locomotor activity recuperation after administration of propofol in Japanese quail (*Coturnix coturnix japonica*). *Research in Veterinary Science.* 93(1):156–162.

Kops, M.S., Kjaer, J.B., Güntürkün, O., Westphal, K.G., Korte-Bouws, G.A., Olivier, B., Bolhuis, J.E., and Korte, S.M. 2014. Serotonin release in the caudal nidopallium of adult laying hens genetically selected for high and low feather pecking behavior: An *in vivo* microdialysis study. *Behavioural Brain Research.* 15(268):81–87.

Lichtenberger, M. 2005a. Determination of indirect blood pressure in the companion bird. *JEPM.* 14(2):149–152.

Lichtenberger, M. 2005b. Normal indirect blood pressure in different species of birds. *Proceedings of the Association of Avian Veterinarians.* 15–17.

Lichtenberger, M., and Ko, J. 2007. Anesthesia and analgesia for small mammals and birds. *Veterinary Clinics of North America: Exotic Animal Practice.* 10(2):293–315.

Ludders, J.W., Mitchell, G.S., and Rode, J. 1990. Minimal anesthetic concentration and cardiopulmonary dose response of isoflurane in ducks. *Veterinary Surgery.* 119(4):304–307.

Lumeij, J.T. 1987. Avian clinical pathology. General considerations. *The Veterinary Quarterly.* 9:249–254.

Lumeij, J.T., and Overduin, L.M. 1990. Plasma chemistry references values in psittaciformes. *Avian Pathology: Journal of the WVPA.* 19:235–244.

Lumeij, J.T., and Remple, J.D. 1991. Plasma urea, creatinine and uric acid concentrations in relation to feeding in peregrine falcons (Falco peregrinus). *Avian Pathology: Journal of the WVPA.* 20:79–83.

Lumeij, J.T., and Remple, J.D. 1992. Plasma bile acid concentrations in response to feeding in peregrine falcons (*Falco peregrinus*). *Avian Diseases.* 36:1060–1062.

Lumeij, J.T., and Westerhof, I. 1987. Blood chemistry for the diagnosis of hepatobiliary disease in birds: A review. *The Veterinary Quarterly.* 9:255–261.

Machin, K.L., and Caulkett, N.A. 1998. Cardiopulmonary effects of propofol and a medetomidine-midazolam-ketamine combination in mallard ducks. *American Journal of Veterinary Research.* 59(5):598–602.

Machin, K.L., and Caulkett, N.A. 2000. Evaluation of isoflurane and propofol anesthesia for intraabdominal transmitter placement in nesting female canvasback ducks. *Journal of Wildlife Diseases.* 36(2):324–334.

Mans, C., Sanchez-Migallon, G.D., Lahner, L., Paul-Murphy, J., and Sladky, K.K. 2012. Sedation and physiologic response to manual restraint after intranasal administration of midazolam in Hispaniolan Amazon parrots (*Amazona ventralis*). *Journal of Avian Medical Surgery.* 26(3):130–139.

Maiti, S.K., Tiwary, R., Vasan, P., and Dutta A. 2006. Xylazine, diazepam and midazolam premedicated anesthesia in white Leghorm Cockerels for typhlectomy. *Journal of the South African Veterinary Association.* 77:12–18.

Mete, A., Stephenson, N., Rogers, K. et al. 2014. Knemidocoptic mange in Wild Golden Eagles, California, USA. *Emergency Infectious Disease.* 20(10):1716–1718.

Moghadasian, M.H. 2002. Experimental atherosclerosis: A historical overview. *Life Science.* 70(8):855–865.

Morishita, T.Y., Johnson, G., and Johnson, G. 2005. Scaly-leg mite infestation associated with digit necrosis in bantam chickens (*Gallus domesticus*). *JAMS.* 19:230–233.

Müller, K., Holzapfel, J., and Brunnberg, L. 2011. Total intravenous anaesthesia by boluses or by continuous rate infusion of propofol in mute swans (*Cygnus olor*). *Veterinary Anaesthesia and Analgesia.* 38(4):286–291.

Muresan, C., Czirjak, G.A., Pap, P.L., and Kobolkuti, L.B. 2008. Ketamine and xylazine anesthesia in the house sparrow. *Bulletin UASVM, Veterinary Medicine.* 65(2):193–195.

Nilson, P.C., Teramitsu, I., and White, S.A. 2005. Caudal thoracic air sac cannulation in zebra finches for isoflurane anesthesia. *Journal of Neuroscience Methods.* 143(2):107–115.

Owiny, J.R., and French, E.D. 2000. Ectoparasites in a pigeon colony. *Comparative Medicine.* 50(2):229–230.

Paré, M., Ludders, J.W., and Erb, H.N. 2013. Association of partial pressure of carbon dioxide in expired gas and arterial blood at three different ventilation states in apneic chickens (*Gallus domesticus*) during air sac insufflation anesthesia. *Veterinary Anaesthesia and Analgesia.* 40(3):245–256.

Paul-Murphy, J.R., Brunson, D.B., and Miletic, V. 1999. Analgesic effects of butorphanol and buprenorphine in conscious African grey parrots (*Psittacus erithacus erithacus* and *Psittacus erithacus timneh*). *American Journal of Veterinary Research.* 60(10): 1218–1221.

Paul-Murphy, J.R., Sladky, K.K., Krugner-Higby, L.A., Stading, B.R., Klauer, J.M., Keuler, N.S., Brown, C.S., and Heath, T.D. 2009. Analgesic effects of carprofen and liposome-encapsulated butorphanol tartrate in Hispaniolan parrots (*Amazona ventralis*) with experimentally induced arthritis. *American Journal of Veterinary Research.* 70(10):1201–1210.

Pence, D.B., Cole, R.A., Brugger, K.E. et al. 1999. Epizootic podoknemidokoptiasis in American robins. *Journal of Wildlife Diseases.* 35:1–7.

Popp, C., Hauck, R., Gad, W., and abd Hafez, H.M. 2012. Type C botulism in a commercial turkey farm: A case report. *Avian Disease.* 56:760–763.

Pritchard, J., Kuster, T., Sparagano, O., and Tomley, F. 2015. Understanding the biology and control of the poultry red mite *Dermanyssus gallinae*: A review. *Avian Pathology.* 21:1–42.

Reim, D.A., and Middleton, C.C. 1995. Use of butorphanol as an anesthetic adjunct in turkeys. *Laboratory Animal Science.* 45(6):696–697.

Reiner, A., Brauth, S.E., Kitt, C.A., and Quirion, R. 1989. Distribution of mu, delta, and kappa opiate receptor types in the forebrain and midbrain of pigeons. *Journal of Comparative Neurology.* 280(3):359–382.

Riggs, S.M., Hawkins, M.G., Craigmill, A.L., Kass, P.H., Stanley, S.D., and Taylor, I.T. 2008. Pharmacokinetics of butorphanol tartrate in red-tailed hawks (*Buteo jamaicensis*) and great horned owls (*Bubo virginianus*). *American Journal of Veterinary Research.* 69(5):596–603.

Ritchie, B.W., Harrison, G.J., and Harrison, L.R. 1994. *Avian Medicine: Principles and Application.* (Lake Worth, Florida: Wingers Publishing, Inc.), p. 148:1077.

Rubinstein, J., and Lightfoot, T. 2014. Feather loss and feather destructive behavior in pet birds. *Veterinary Clinics of North America: Exotic Animal Practice.* 17(1):77–101.

Ryan, C.P. 1992. Avocado poisoning. *Journal of the American Veterinary Medical Association.* 2000(12):1780.

Sandmeier, P. 2000. Evaluation of medetomidine for short-term immobilization of domestic pigeons (*Columba livia*) and Amazon parrots (*Amazona species*). *JAMS.* 14(1):8–14.

Sladky, K.K., Krugner-Higby, L., Meek-Walker, E., Heath, T.D., and Paul-Murphy, J. 2006. Serum concentrations and analgesic effects of liposome-encapsulated and standard butorphanol tartrate in parrots. *American Journal of Veterinary Research.* 67(5):775–781.

Siddalls, M., Currier, T.A., Pang, J., Lertpiriyapong, K., and Patterson, M.M. 2015. Infestation of research zebra finch colony with 2 novel mite species. *Comparative Medicine.* 65(1):51–53.

Silim, A., Thorsen, J., and Carlson, H.C. 1978. Experimental infection of chickens with hemorrhagic enteritis virus. *Avian Diseases.* 22(1):106–114.

Shuster, K.A., Brock, K.L., Dysko, R.C., DiRita, V.J., and Bergin, I.L. 2012. Polytetrafluorocthylene toxicosis in recently hatched chickens (*Gallus domesticus*). *Comparative Medicine.* 62(1):49–52.

Singh, P.M., Johnson, C., Gartrell, B., Mitchinson, S., and Chambers, P. 2011. Pharmacokinetics of butorphanol in broiler chickens. *Veterinary Record.* 168(22):588.

Snyder, J.M., Molk, D.M., and Treuting, P.M. 2013. Increased mortality in a colony of zebra finches exposed to continuous light. *JAALAS.* 52(3):301–307.

Toparlak, M., Tuzer, E., Gargili, A., and Gulabner, A. 1999. Therapy of Knemidocoptic mange in budgerigars with spot-on application of moxidectin. *Turkish Journal of Veterinary and Animal Sciences.* 23:173–174.

Uzun, M., Onder, F., Atalan, G., Cenesiz, M., Kaya, M., and Yildiz, S. 2006. Effects of xylazine, medetomidine, detomidine, and diazepam on sedation, heart and respiratory rates, and cloacal temperature in rock partridges (*Alectoris graeca*). *Journal of Zoo and Wildlife Medicine.* 37(2):135–140.

Valverde, A., Honeyman, V.L., Dyson, D.H., and Valliant, A.E. 1990. Determination of a sedative dose and influence of midazolam on cardiopulmonary function in Canada geese. *American Journal of Veterinary Research.* 51(7):1071–1074.

Van Meir, V., Boumans, T., De Groof, G., Van Audekerke, J., Smolders, A., Scheunders, P., Sijbers, J., Verhoye, M., Balthazart, J., and Van der Linden A. 2005. Spatiotemporal properties of the BOLD response in the songbirds' auditory circuit during a variety of listening tasks. *Neuroimage.* 25(4):1242–1255.

Zhao, K., He, W., Xie, S., Song, D., Lu, H., Pan, W., Zhou, P. et al. 2014. Highly pathogenic fowlpox virus in cutaneously infected chickens, China. *Emerging Infectious Disease.* 20(7):1208–1210.

experimental methodology

This chapter addresses many of the experimental techniques commonly used when working with avian species in a research setting. Advanced techniques such as surgical procedures or operant conditioning are beyond the scope of this work. Since species variations may affect the feasibility or methods for some procedures, the reader is advised to review the unique biology and behavior of the avian species at hand prior to initiating work. Additionally, proper training of personnel employing these techniques is essential to ensure human and animal safety and welfare.

capture and restraint

Most procedures require capture and restraint of avian subjects, which can be stressful to both the bird and the handler when done improperly. Passerines and wild birds are especially susceptible to stress associated with capture, and some species of birds (e.g., psittacines and raptors) are capable of causing significant injury to personnel if handled incorrectly. The species of bird dictates the specific method of capture and restraint.

In order to minimize risk to the avian subject and the handler, personnel should be appropriately trained, and restraint time should be minimized to reduce patient overheating and distress. Since birds are naturally fearful of humans, it is often beneficial to acclimate birds to the technicians that will be working with them. There is evidence that early handling and habituation may reduce fearfulness in some species (Hughes and Black, 1976; Fox and Millam 2004).

Some general guidelines for successful avian capture and restraint include the following:

- Close any doors to prevent escape from the housing room prior to opening the primary enclosure and minimize hiding places (e.g., under carts and on top of shelves). If necessary, remove items from the bird's cage that may complicate capture.

- Dimming the lights in the room or darkening the cage often calms excitable subjects. A penlight or small flashlight can be used if necessary to locate the bird.

- Never catch birds by the wings or legs alone as this may cause injury.

- During restraint, keel movement must be unrestricted to allow adequate thoracic motion for respiration.

- Keep the bird upright whenever possible to avoid compromising airflow through the respiratory system.

- With prolonged restraint (>5 min), some subjects may become excessively stressed and overheat. Minimize handing time to what is absolutely necessary, and if prolonged, monitor for increased respiratory rate. The respiratory rate rises as a subject's temperature rises (Cabanac and Aizawa, 2000; Geenacre and Lusby, 2004).

- Baby chicks may become hypothermic if removed from supplemental heat sources for too long.

- Release birds onto the floor or cage bottom facing away from the handler and any obstructions that might cause injury should the animal panic. Do not release birds in midair (Hawkins et al., 2001).

- After release, birds should return to normal activity within a few minutes.

- For transport between rooms, a transport cage or carrier is recommended to prevent escape.

capture techniques

Avian capture can be challenging for the inexperienced individual and represents a point at which the animal or handler is susceptible to injury. Improper capture technique resulting in stress may also invoke fear in the subject during subsequent capture attempts.

Unless birds are habituated to handling, calm, quick, deliberate movements are necessary to prevent panic flights or fear-induced aggression.

The type of capture equipment required will depend on the species of bird and the housing arrangement. Most songbirds, domestic fowl, and ducks can be caught using bare hands. Some personnel prefer to use a paper towel or small piece of surgical drape to capture small birds. Cloth towels are most commonly used to restrain psittacines and can be useful for concealing the catching hand, obstructing the bird's vision, and preventing wing flapping. Towels must be free of frayed edges and hanging threads as these may become wrapped around the subject's feet. Nets are primarily used to capture birds in pens and large flight cages, and birds that have escaped from the primary enclosure. A small gauge mesh is preferred to minimize the risk of leg and wing entanglement. Leather gloves, though commonly used in falconry, are not generally recommended as they restrict movement and can injure the bird.

General recommendations for capturing commonly used species are as follows:

- Caged songbirds are caught by quickly cupping the hand around the back and neck of the bird. This is most easily accomplished when the bird is grasping the cage bars. Enough pressure should be applied to prevent the wings from flapping, but not so much as to restrict thoracic movement.

- Psittacines are caught by first grasping around the back of head with the thumb and index finger on either side of the lower jaw. This is easiest with the bird in a corner or climbing on the cage. A towel is useful to initially grasp the bird, and then the towel can be wrapped around the bird's body to secure the wings.

- Pigeons, domestic fowl, and ducks can be caught with two hands, pinning the wings firmly against the bird's sides so they cannot flap freely. Birds should be approached slowly from behind and quickly grasped.

- Group-housed birds in pens tend to flock and can be walked into a corner for capture by hand or with a net (Hawkins et al., 2001).

- Nets can be dropped over the bird for capture. Once netted, the bird's head and wings must be restrained before removal from the net to avoid injury.

manual restraint

Once captured, birds may be manually restrained for examinations or noninvasive experimental procedures. General recommendations for manual restraint of common lab species are as follows:

- Passerines are restrained by placing the head between the handler's index and middle finger and cradling the back of the bird in the palm of the hand. The thumb and remaining fingers are gently encircled around the bird to secure the wings and legs (Figure 5.1).

- To restrain psittacines, the lower jaw, wings, and feet must be immobilized. The head of small birds (<150 g) is restrained by placing the thumb and middle finger on opposite sides of the bird's cheek and the index finger over the back of the head (Wade, 2009). The head of larger psittacines is restrained by wrapping the fingers under the mandible, keeping the neck stretched (Wade, 2009). The handler's second hand is then used to control the lower wings and legs.

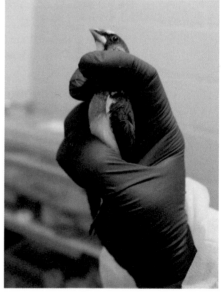

Fig. 5.1 Manual restraint of a Bengalese finch. Note the bird's head is between the handler's index and middle fingers, and the breast is not constricted to allow adequate respiration.

- Poultry and small waterfowl restraint is similar to carrying a football. With the head facing forward, the bird is grasped and held against the handler's body. The handler's free arm slides under the bird's belly to support the body, and the legs are restrained between the fingers of the same hand. The other hand is placed over the back to restrain the outer wing (Hawkins et al., 2001) (Figure 5.2).

- Large geese, which are capable of biting, are restrained by using one arm to secure the bird against the handler's side and support the body, and the other hand to secure the neck.

- Chicks and ducklings can be carried in the palm of one hand with the fingers under the body (Hawkins et al., 2001).

- Pigeons and doves are restrained using a one- or two-handed technique. Small birds can be held in one hand with the legs immobilized between the index and middle finger and the end of the wings encircled by the thumb and remaining fingers. Larger birds are restrained using two hands; the second hand is used to support the front of the wings and breast (Figure 5.3).

- A towel wrap is commonly used to limit wing movement in medium- to large-sized birds but can cause overheating if kept in place for more than a few minutes. A pictorial description of towel restraint can be found in Kalmar et al. (2007).

- A cloth head cover may calm subjects during procedures. Restraint tubes are also described (DellaVolpe et al., 2011).

Fig. 5.2 Manual restraint of a chicken.

Fig. 5.3 Two-handed technique for manually restraining a pigeon. (Drawing courtesy of Kristy Roper. With permission.)

chemical restraint

Chemical restraint is required for potentially painful procedures, in especially frightened birds, or for complete immobilization to ensure safe and adequate sample collection. Chemical restraint should be considered for prolonged procedures due to the risk of stress-induced hyperthermia or shock. Passerines in particular are prone to significant and sometimes fatal effects of stress. The reader is referred to Chapter 4: Veterinary Care for specific information on the use of chemical restraint agents in birds.

various sampling techniques

Many sampling procedures can be conducted in physically restrained birds except where noted. Some birds can be trained to cooperate with common, noninvasive procedures such as weighing, medication administration, or palpation using operant conditioning.

Oropharyngeal and Cloacal Swabs

Oropharyngeal and cloacal swabs are often used in pathogen isolation during infectious disease studies.

Procedure:

1. Appropriately restrain the bird.
2. Moisten a cotton-tipped swab/culturette with saline or transport medium.

3. To collect an oropharyngeal swab, open the bird's mouth and roll the swab around the trachea and up along the choanal slit. A mouth speculum or strips of gauze around the upper and lower beak can be used to keep the mouth open.

4. To collect a cloacal swab, part the feathers around the vent and gently insert the swab into the cloaca using a twisting motion. Once the swab is completely inserted, rub along the mucosal wall and remove.

Sinus Flush

Microbes and cellular contents of the nasal passages and sinuses can be obtained for evaluation via a nasal flush. Before performing this procedure, always palpate the bird's crop and ensure it is empty to reduce the risk of aspiration.

Procedure:

1. Fill a syringe with a small amount of saline appropriate for the size of the bird (e.g., 1–3 mL for small birds) and remove the needle.

2. Hold the bird in an inverted position with the head lower than the body, occluding the esophagus to prevent aspiration of any crop contents.

3. Administer the saline into the sinus cavity via the nares; communication normally exists between both nares and the choana.

4. Collect the sample into a tube or aspirate from a clean surface.

Crop Lavage

A crop lavage is performed to obtain contents for evaluation. The procedure may be necessary in ecology or nutrition studies.

Procedure:

1. Appropriately restrain the bird. One person is usually required to restrain the bird while the second person performs the lavage.

2. The procedure is performed by placing a ball-tipped gavage needle or flexible tube attached to a syringe in the crop. This is easily accomplished by simply avoiding the glottis. The syringe may be filled with a small amount of water if the crop is empty.

3. Tube placement in the crop is the same as for crop feeding (see Oral Administration: Crop).

4. After confirming placement, inject water into the crop. Aspirate the contents back into the syringe, being careful not to suction against the crop and cause bruising.

5. The contents can be used for microbial culture, cytological evaluation, or other experimental evaluations.

Semen Collection

Semen collection may be used in studies of reproduction. The abdominal massage technique is commonly used in domestic fowl and has been described in finches, ducks, pigeons, doves, quail, and waterfowl (Gee et al., 2004). This is a two-person procedure in which the bird is positioned with the head downward and the tail toward the other person. The collector uses a series of stroking motions on the abdomen and cloaca to stimulate cloacal eversion and ejaculation. Minor modifications are made for different species.

Tissue Biopsies

Antemortem tissue samples are often required for physiology, disease, reproduction, and genetic studies (Fair et al., 2010). A surgical approach is traditionally used for internal organ biopsies necessitating chemical restraint, analgesia, and aseptic technique. Increasingly, tissue biopsies are obtained endoscopically using a miniature rigid endoscope. Although specialized equipment and skill are required, endoscopic collection has several advantages. Endoscopic biopsies are minimally invasive, thus reducing blood loss, patient discomfort, and recovery time. Additionally, many endoscopic procedures are considered minor surgery by Institutional Animal Care and Use Committees (IACUCs), making serial tissue sampling possible. For more information on common approaches, the reader is referred to the work by Divers (2010).

Urine/Feces Collection

Urine or feces can be collected from a fresh voided sample on a clean, water-resistant surface. Urine is separated by aspirating the liquid with a needle and syringe or a pipette, or via the capillary action of a microcapillary tube. Because feces and urine are voided through a single orifice, the cloaca, there is inevitably some mixing of the two.

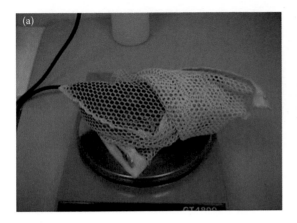

Fig. 5.4 (a) A well-ventilated cloth bag is used to contain a bird for weight measurement. (b) Disposable food takeout containers are useful for weighing small passerines.

To obtain a clean urine sample, cloacal cannulation is a viable option that has been described in pigeons (Halsema et al., 1988).

Body Weight Measurement

Weight measurement is routinely necessary to determine drug doses, safe blood collection volumes, and to monitor overall health. Depending on the size of the bird, a scale capable of measuring weight in grams or kilograms is required. The bird is caught and can be placed in a cloth bag or a container that allows sufficient ventilation (Figure 5.4a). Chinese-style food take-out boxes work well for small passerines, and plastic bins can be used for larger birds (Figure 5.4b). The scale must be tared for the weight of the container. For studies in which frequent weight measurement is required, operant conditioning to allow capture and weighing or to train the bird to stand on a scale should be considered.

blood collection

Avian blood volume varies by species, ranging from 5% to 20% of body weight (Campbell, 1994). As a general rule, blood volume is calculated as approximately 10% of body weight. Healthy birds can tolerate loss of up to 10% of their blood volume, or the equivalent of 1% body weight in kilograms, without detriment. Using this guideline, 1 mL can be safely taken from a 100 g bird in a single collection unless the bird is dehydrated or anemic. In small species, an accurate body weight and blood volume calculation is critical prior to performing blood collection. For these birds, a few extra drops of blood loss may constitute a significant percentage of the subject's blood volume. Even blood loss into a hematoma can cause harm to small passerines. For all blood sample collections, care should be taken to collect the smallest volume necessary to meet the needs of the study.

Blood is collected using a needle and syringe, the size of which is dictated by the size of the subject and the vessel. As a general rule, the diameter of the needle should be just smaller than the diameter of the vessel to reduce the risk of hemolysis and facilitate rapid collection without lacerating the vein. A short 22–28 g needle and 1–3 mL syringe is appropriate for most birds. Butterfly needles are useful when needle movement, due to either the bird or an unsteady hand, is a concern. During blood withdrawal, slow and steady aspiration is required to avoid collapsing the vessel. For small samples, an unattached needle can be used to prick or insert in the vein, and blood is allowed to drip into microcollection device or a capillary tube. This method may be useful in very small birds.

After collection, blood samples should be immediately transferred into sample vials for processing. For hematology studies, ethylenediaminetetraacetic acid (EDTA) (typically in "purple top" tubes) is the anticoagulant of choice although it is important to note that it may cause hemolysis of erythrocytes in Corvids (Campbell, 1994). For serum chemistry, lithium heparin coated ("green top") or plain tubes ("red top") are preferred. Practically speaking, vials containing lithium heparin are acceptable for both hematology and chemistry analysis making this a reasonable default choice for most avian samples. Avian blood does not generally preserve well; so, it is recommended that samples be processed as soon as possible after collection.

Blood smears can be made with the standard two slide push technique or by placing a drop of blood on a slide, placing a coverslip on top and pulling the coverslip off when the blood begins to spread. The blood smear should be made immediately after collection. A variety of stains can be used including Wright's, Giemsa, Wright's Giemsa, and Romanowski stains. If blood is being evaluated for parasites, it is recommended that an anticoagulant not be used (Campbell, 1994). See Chapter 4: Veterinary Care for additional details about the hematology of avian patients.

blood collection sites

The choice of sampling site depends on the species and size of the bird, and the volume of blood required for analysis. A common site for blood collection in many avian species is the jugular vein, and it may be the only site for a survival collection in smaller species. Relatively large volumes can be collected from the jugular vein. Other sites in medium to large species include the medial metatarsal vein, and the ulnar (wing) vein. Toe clipping and phlebotomy from the occipital sinus, although used historically, are not recommended as survival procedures due to the potential for residual pain and skewed cell distributions (toe clipping), and life-threatening complications (occipital sinus) (Campbell, 1994). Terminal procedures useful for large blood collections include cardiocentesis under deep anesthesia and decapitation. For studies that require repeated blood samples, placement of an indwelling catheter (short-term) or vascular access port (long-term) will facilitate collection (Figure 5.5). Intravenous and intraosseus catheter placement are described under the section titled, Compound Administration.

Although most collections can be conducted using physical restraint, chemical restraint may be required for adequate immobilization or to alleviate stress. The benefits of either method should be weighed against the potential risks. Unless the bird is anesthetized, two people are generally required for the procedure, although blood can be collected from the jugular vein of small species by a skilled operator.

Fig. 5.5 Intra-arterial catheter placement for sampling blood gases in a pigeon.

Survival Collection Techniques

Jugular vein

The right jugular vein is preferred over the left due to its larger size in many avian species (Figure 5.6). Hematomas are common at this site if personnel are not careful to ensure adequate hemostasis or the bird is not adequately restrained. Jugular vein blood collection is not recommended in species with a plexus venosus subcutaneous collaris (e.g., pigeons).

Procedure:

1. Place the bird in dorsal or left lateral recumbency.
2. The head and neck are gently secured in one hand (generally the left hand if the handler is right-handed) and gentle traction is applied to extend the neck.
3. Lightly wet the feathers with alcohol over the vein; however, there is often a featherless tract here. The vein should be visible through the skin.
4. Use a finger to apply pressure to the vein just cranial to the thoracic inlet. This will occlude and distend the vein making it easier to collect the sample.
5. Insert the needle with the bevel up into the vein at a slight 20–30° angle. The needle should point toward the head.
6. As soon as the needle penetrates the vein, gently withdraw the syringe plunger. The syringe should begin filling with blood. If blood does not flow, slightly readjust the position of the needle.

Fig. 5.6 Blood collection from the jugular vein.

7. Slowly aspirate the sample, taking care not to collapse the vessel. If hematoma formation is observed, remove the needle and apply pressure to the site.

8. Once the collection is complete, withdraw the needle and immediately apply pressure to the site for at least 1 min to ensure hemostasis.

Ulnaris (wing) vein

This site is commonly used in poultry medicine and can be used for blood collection in medium- to large-sized birds (Figure 5.7). It is not recommended in small birds due to the high potential of vessel laceration and hematoma formation. Chemical restraint may facilitate collection and help prevent complication as hematoma formation in birds of all sizes is common due to the lack of surrounding subcutaneous tissue.

Procedure:

1. Typically, two people are required for this method. The bird is restrained in dorsal recumbency by the first person. For chickens, one hand is used to hold the legs and the other

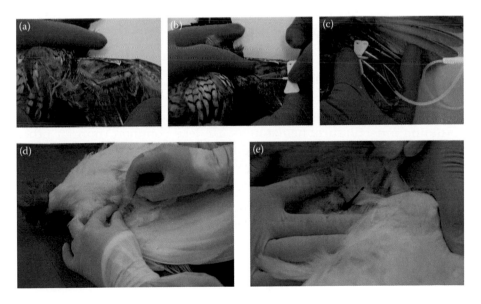

Fig. 5.7 (a–c) Blood collection from the ulnar vein using a butterfly needle. (d, e) Alternatively, blood is obtained by pricking the ulnar vessel with a needle and collecting the sample in a capillary tube. (Photo courtesy of Leanne Alworth. With permission.)

hand to restrain the breast and the wing. The second person extends the opposite wing for sampling.

2. Wet the feathers over the vein lightly with alcohol. If necessary, pluck any feathers obscuring the site.

3. Identify the vein where it crosses ventral to the humero-radioulnar joint (elbow). The assistant applies pressure to the vein over the humerus. This will occlude and distend the vein making it easier to collect the sample.

4. Insert the needle with the bevel up into the vein at a slight 20–30° angle. The needle should point toward the heart.

5. As soon as the needle penetrates the vein, gently withdraw the syringe plunger. The syringe should begin filling with blood. Alternatively, if a syringe is not used, the needle hub should immediately begin filling with blood. If blood does not flow, slightly readjust the position of the needle.

6. Slowly aspirate the sample taking care not to collapse the vessel. If a syringe is not used, allow the blood to drip into a microcollection tube. If hematoma formation is observed, remove the needle and apply pressure to the site.

7. Once the collection is complete, withdraw the needle and immediately apply pressure to the site for at least 1 min to ensure hemostasis.

Medial metatarsal vein

This site can be used in most birds except very small ones and is commonly used in waterfowl or larger psittacines. Due to the surrounding musculature, hematomas are less common than for other methods. This vein may be more difficult to identify than the jugular or ulnar vein. Smaller volumes are collected from this site.

Procedure:

1. Typically, two people are required for this technique. The bird is restrained in lateral or dorsal recumbency by the first person. The second person extends the leg for sampling.

2. Lightly wet the area over the vein with alcohol. If necessary, pluck any feathers obscuring the site.

3. Identify the vein as it runs along the medial side of the tarsal-metatarsus at the tibiotarsal-tarsometatarsal joint (knee). The assistant applies pressure to the vein over the knee. This

will occlude and distend the vein making it easier to collect the sample.

4. Follow procedure steps 4–7 as for ulnaris vein collection.

Terminal (Nonsurvival) Techniques

Occipital venous sinus

Although this site has been used historically for repeatable, survival collections (Zimmermann and Dhillon, 1985), the authors recommend this technique be reserved only for terminal collections in anesthetized birds due to the potential for brainstem injury. Relatively large volumes can be collected from this site. Needles affixed to a vacuum blood collection tube are required.

Procedure:

1. Anesthetize the bird.
2. Restrain the bird so that the head forms a straight line with the spine and flex the head ventrally.
3. Locate the occipital venous sinus by palpating the depression between the base of the skull and the first cervical vertebrae. Lightly wipe the area with alcohol.
4. Direct the needle of appropriate size for the bird at a 30–40° angle to the cervical vertebrae on the dorsal midline.
5. Once the skin is penetrated, push the collection tube on the rubber end of the needle and advance the needle slightly. If the needle is properly placed, blood should rush into the tube. Do not remove the needle from the skin without removing the tube or the vacuum will be lost.
6. Once the sample is collected, remove the tube from the vacutainer needle and withdraw the needle from the skin. Apply pressure at the site.

Cardiocentesis

Although this technique produces the largest yield of blood, it should only be used in anesthetized birds prior to euthanasia due to the potential for pain and fatal hemorrhage. The choice of needle and syringe will depend on the size of the bird. A longer needle (>1 in) may be required in large birds. The heart can be penetrated from an anterior approach through the thoracic inlet, from a ventral approach directing the needle underneath the sternum and last ribs, or from a lateral approach through the fourth intercostal space. The approach of choice will vary by species. Galliformes are typically bled from the lateral approach. Once the needle is inserted into the heart, the syringe plunger is withdrawn and the sample collected similar to peripheral collection.

compound administration

Experimental and therapeutic compounds can be administered via parenteral and oral routes in birds. Parenteral administration allows compounds to enter the vascular system more directly and is generally more precise than oral administration. Common routes for parenteral administration in birds include subcutaneous (SC), intravenous (IV), intraosseus (IO), and intramuscular (IM). Although sometimes used to administer euthanasia agents, intracoelomic injections are generally discouraged due to the risk of accidental administration into the air sacs.

The route of drug administration is affected by factors such as pH, volume, absorption, frequency of injection, species characteristics, number of birds to be treated, and skill of personnel (Morton et al., 2001). Avian formularies or a review of the literature can help guide the optimal route of entry for a particular compound. However, since many pharmaceuticals administered to birds have not been formally investigated in avian species and are given off-label, the dose and route of administration must often be extrapolated from other species.

Parenteral Techniques

Subcutaneous (SC) administration

Subcutaneous administration (SC) is typically used for fluid therapy, or when intravenous or intramuscular injections are not feasible or are contraindicated. Compounds are more slowly absorbed by this route and pooling in the subcutaneous space may be observed with larger volumes. Subcutaneous injections can be given in the unfeathered inguinal web of skin between the leg and body wall, the wing web at the axilla, and in the interscapular region (Figure 5.8). The inguinal region is generally best for larger volumes, whereas the interscapular region is preferred for stressed birds or birds requiring minimal restraint. Care must be taken when injecting in the interscapular area to avoid accidental injection in the cervical air sac located at the base of the neck (Wade, 2009).

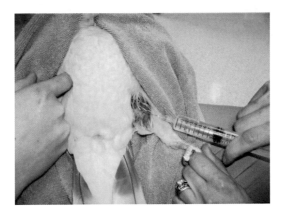

Fig. 5.8 Subcutaneous fluid administration in the inguinal web of a parrot.

Procedure:

1. The needle gauge is guided by the size of the bird and the viscosity of the compound to be administered. The larger the needle, the easier the injection will be but the more likely it is for fluid to leak from the injection site.

2. Appropriately restrain the bird. One person may be required to hold the bird while the other gives the injection.

3. For inguinal web injections, pull the leg toward you and identify the web of skin between the thigh and the body wall. For wing web injections, extend the wing and look for the web of skin between the wing and the body wall. Interscapular injections are given on the back between the elbows (Wade, 2009).

4. If possible, lightly swipe the injection site with alcohol. If feathers are present, an alcohol swab may be used to separate the feathers.

5. For inguinal and wing web injections, insert the needle with attached syringe into the skin web, taking care to avoid penetrating through the other side. For interscapular injections, gently lift the skin to create a fold at the injection site and insert the needle into the base of the fold.

6. Slowly administer the injection. If the needle is correctly placed a bleb of fluid will form under the skin.

7. The maximum volume for injection is guided by the size of the bird and the location. Large volumes (>5 mL/kg) should be administered between several sites.

Intravenous (IV) administration

Intravenous administration (IV) is used for direct delivery of a compound into the peripheral circulation or for fluid therapy. The maximum plasma concentration of drug given IV is generally more rapidly attained than when given IM, SC, or oral (Morton et al., 2001). Common sites for IV injection are generally those that are also used in blood collection and include the right jugular vein, ulnar (wing) vein, and medial metatarsal vein. Since avian veins are fragile, larger volumes are best administered in the jugular vein. Exceptions to this are members of the *Columbiformes* family in which jugular injections are contraindicated due to the presence of the plexus venosus subcutaneous collaris.

IV injections are generally appropriate for a single administration due to the risk of hematoma formation. If multiple injections are required over a short interval, either the vessels should be alternated or an indwelling IV catheter should be placed. Some species, however, especially psittacines, will not readily tolerate IV catheters and inadvertent removal could cause fatal hemorrhage. Intraosseous catheters are an alternative that are more stable and can be used when longer term (up to 72 h) access to the vascular system is required. For chronic administration, surgically implanted vascular access ports should be considered.

The protocol for IV injections is essentially the same as the procedure for blood collection (see the section titled, Blood Collection). As for blood collection, small gauge butterfly needles will reduce hematoma formation by allowing the needle to move if the subject moves. Although IV injections can be conducted under physical restraint, chemical restraint may be required to immobilize the patient or for highly stressed birds. Once the needle is inserted, the syringe plunger should be withdrawn to determine that the needle is in the vein. It is important to inject compounds slowly to avoid damaging the vein. If leakage of the drug around the vein or hematoma formation is observed, the injection should be stopped, the needle removed, and pressure applied to the site.

IV Catheter Placement Technique

1. Catheters can be placed in the right jugular, ulnar, and medial metatarsal vein. The gauge of the catheter is determined by the size of the vessel. A 24 g catheter may be used in small vessels, whereas a 20–22 g is used in larger vessels.

The jugular vein is easiest to catheterize and can be used for smaller birds; however, this site is more difficult to secure. Ulnar veins may be difficult to catheterize because of the smaller access area.

2. Appropriately restrain the bird. Chemical restraint is often required.

3. Pluck feathers if necessary and aseptically prepare the site.

4. Visualize the vein and apply pressure cranial to the insertion site. Having an assistant to do this is helpful.

5. Insert the catheter at a 30° angle with the bevel up in the direction of the heart.

6. Once blood is visualized in the hub, advance the catheter into the vein while holding the stylet steady. If the catheter if properly placed, it will advance easily. If resistance is encountered, the catheter is likely not in the vein lumen and a placement should be cautiously reattempted.

7. Completely remove the stylet and attach an injection cap to the catheter.

8. Flush the catheter with heparinized saline through the injection cap to ensure the catheter is patent and there is no leakage around the vessel. If resistance is encountered or extravasation is observed, remove the catheter and apply pressure to the site.

9. If the catheter will be maintained beyond the initial injection, secure the catheter by attaching a tape butterfly and suturing the tape to the skin, or wrapping in a bandage. Jugular and leg catheters can be secured by wrapping the catheter with a bandage around the neck or leg, respectively. Ulnar catheters can be stabilized by affixing it with tape to a tongue depressor that extends past the end of the catheter, and incorporating the proximal and distal ends of the tongue depressor in an elastic bandage wrapped around the humerus and ulna in a figure eight pattern.

Intraosseus (IO) administration

Intraosseus administration involves injecting a compound through a catheter directly into the medullary canal of a bone. It is used when repeated drug or fluid therapy is necessary and is considered equivalent to IV administration in terms of absorption rate (Quesenberry and Hillyer, 1994). Chemical restraint is required for this technique to allow patient positioning and minimize discomfort. Since IO injections can be painful, this route should be reserved for emergency care or when other routes are not feasible.

IO injections can theoretically be given in any marrow-containing bone; pneumatic bones (e.g., humerus and femur) must be avoided to prevent injection into the respiratory system. Common sites for IO injection include the proximal tibiotarsus and the distal ulna. With aseptic placement and proper care, IO catheters can remain in place for up to 72 h (Quesenberry and Hillyer, 1994). Note that bone marrow samples can be collected for cytology during initial placement of an IO catheter.

Procedure:

1. A short spinal needle (18–22 g) can be used for larger birds, whereas a small hypodermic needle (22–25 g) can be used in very small birds. The size of the bone guides the size and length of the needle. Since needle placement involves drilling through bone, spinal needles are preferred when possible. The stylet prevents bone from clogging the needle lumen.

2. Chemically restrain the bird.

3. Remove feathers and aseptically prepare the insertion site. Sterility is extremely important because contamination can result in bacterial osteomyelitis.

4. a. *Placement in the distal ulna*: Palpate the crest of the distal ulna at the carpus. The thumb of one hand is placed along the ulna to use as a guide for positioning the needle. Insert the needle between the distal ulna and the ulnar carpal bone by twisting the needle with firm pressure until it enters the medullary canal and the needle can be advanced. Aspiration should yield bone marrow or blood.

   b. *Placement in the proximal tibiotarsus*: Palpate the tibial crest at the cranial proximal portion of the tibiotarsus. Insert the needle through the tibial crest by firmly twisting the needle and pass distally into the medullary cavity. Aspiration should yield bone marrow or blood.

5. Attach a small syringe and slowly flush the catheter with heparinized saline. If properly placed in the distal ulna, the ulnar vein should blanch. There is no similar vein to observe in the proximal tibiotarsus. If strong resistance is encountered, or

if fluid is seen under the skin surrounding the catheter, then the needle is not properly placed within the medullary cavity. Alternatively, a radiograph can be used to confirm placement.

6. Once patency is confirmed, remove the syringe and attach an injection cap to the end of the catheter.

7. Administer the compound slowly through the injection port. Since the medullary cavity cannot expand, some resistance is expected; however, the drug should flow freely through the catheter. Do not use excessive force to inject.

8. If the catheter is kept beyond the first injection, it must be secured to the wing or leg. Ulnar catheters are secured by applying sterile tissue adhesive at the point of insertion and wrapping the wing around the humerus and ulna with an elastic bandage in a figure eight pattern. Leg catheters are similarly secured wrapping an elastic bandage around the leg.

9. Catheters may remain in place for up to 72 h if flushed with heparinized saline every 6–8 h.

Intramuscular (IM) administration

This is a common route of administration in avian species and absorption from this site is more rapid than the subcutaneous route. IM injections are most frequently given in the pectoral muscles because of their large size (see Chapter 1: Important Biological Features). For species with less developed pectoral muscles, the thigh muscles may be used if large enough to accommodate the injection volume; however, drugs administered in this location may pass through the renal portal system and be eliminated prior to entering the general circulation (Flammer, 1994). For this reason, drugs that are nephrotoxic should never be injected in the caudal half of the body.

Some compounds may cause muscle necrosis when given IM. In order to reduce this risk, it is recommended to use a small short needle, alternate injection sites when giving repeated injections, and divide larger volumes between several sites to minimize muscle damage. Additionally, compounds that are known to be irritating should not be given IM.

The maximum volume for IM injection will vary by species, but as a general guideline, no more than 0.1 mL should be administered in a single site in finches, and no more than 1.0 mL in macaws (Schulte and Rupley, 2004).

Procedure:

1. Use the smallest gauge needle that allows the administered fluid to flow. A short 30 g needle is appropriate for very small birds, whereas a 22–25 g needle is suitable for larger birds.
2. Appropriately restrain the bird. Physical restraint is typically sufficient.
3. Palpate the muscle group to be injected. IM injections should be given in a thick part of the muscle. For pectoral muscles, this is usually close and perpendicular to the keel. Directing the needle toward the keel greatly reduces the risk of inadvertently penetrating lungs or air sacs.
4. If possible, lightly swipe the injection site with alcohol.
5. Insert the needle a few millimeters into the muscle. Slightly withdraw the plunger to ensure the needle is not in a blood vessel. If it is clear, inject the compound. If blood is seen in the hub, withdraw the needle and retry at a different location.

Vascular access ports (VAPs)

A vascular access port is a surgically implanted IV catheter connected to a subcutaneous access port that allows chronic blood sampling or compound administration without the need to repeatedly puncture a vessel. Since the entire system is enclosed under the skin, there is minimal risk of the bird removing the device. Injections are given into the subcutaneous port using a noncoring Huber needle designed to preserve the integrity of the port's injection window. VAP use in several bird species has been described (Harvey-Clark, 1990; Senthilkumaran et al., 2006).

This technique may be considered a refinement in studies where serial sampling is required. However, VAPs may only be possible in larger avian subjects, and the need for surgical implantation should be weighed against the potential benefit of the device. Additionally, since VAPs connect directly to the peripheral circulation, there is a risk of septicemia if strict aseptic technique is not followed during access. VAPs must be maintained with regular saline flushing and administration of locking solution to prevent thrombosis. With proper care, they can remain patent for long periods of time.

Implants

For studies where long-term (up to several weeks) continuous drug delivery is required, osmotic mini-pumps or slow-release pellets can be used. The devices are commonly placed in the subcutaneous tissue on the flank and sides of the thorax (Fair et al., 2010). Insertion requires a minor surgery using aseptic technique.

Enteral Techniques

Oral administration

Delivery of a compound in food or water is the least stressful method of administration since capture and restraint are not required (Morton et al., 2001). Although this technique is quite easy and may be the only practical method of drug delivery in large flocks, it is generally not recommended for experimental studies due to variations in individual consumption that can lead to inaccurate dosing. If precise dosing is required, a different method of administration is necessary. Other points for general consideration include

- All other sources of food or water must be removed to ensure only the medicated water/diet is consumed.
- Additional time and effort may be required by husbandry personnel.
- Medicated water is generally unpalatable and birds may refuse to drink. Food-based medications can be delivered in a preferred food item, however, some birds may still fail to consume adequate amounts or refuse the item altogether. Additionally, there is a risk of nutritional imbalance in birds consuming only preferred items over a long period of time.
- Food or water can become contaminated with droppings and must be replaced, resulting in drug wastage.
- The drug must be homogeneously distributed in the food or water. Compounds delivered in the water must be water soluble. Some medications can be commercially compounded into the food to ensure an even distribution.
- The stability of the compound must be taken into account. Some medications break down very rapidly at housing room temperatures.

Oropharyngeal

Compounds may be delivered directly into the gastrointestinal tract by either oropharyngeal administration or oral gavage (Figure 5.9). Orophyngeal administration permits dosing that is more accurate

Fig. 5.9 Oropharyngeal administration in a chicken. (Photo courtesy of Leanne Alworth. With permission.)

than voluntary oral consumption and is relatively easy to perform, but it carries a higher risk of aspiration and may result in some spillage of the product. Only small volumes of substances (<1 mL in chickens) should be given via this route (Alworth and Kelly, 2014). To conduct the procedure, one person is required to restrain the bird while a second person opens the beak. The compound is administered in the far right side of the mouth behind the glottis using a catheter-tipped syringe. Care must be taken to ensure the syringe is not placed directly in the esophagus as this will induce trauma. Birds such as pigeons and chickens that are amenable can similarly be administered oral capsules (Flammer, 1994).

Crop

Crop feeding is used to introduce compounds directly into the crop. This is the most precise method of enteral dosing but requires the utmost skill to be performed safely. The technique used is similar to that used for crop lavage described earlier in the chapter. Curved, ball-tipped gavage needles, similar to those used in rodents, are recommended for gavage in smaller species (Figure 5.10) (Lennox, 2006). A flexible red rubber catheter is suitable for larger birds although care must be taken to ensure the bird does not bite the tube (Figure 5.11). An oral speculum, or alternatively gauze strips placed on the upper and lower beak to hold open the mouth, must be used to keep the mouth open. This is not necessary in species such as ducks that are incapable of lacerating the tube (Lennox, 2006).

Fig. 5.10 Oral administration using a ball-tipped gavage needle.

Procedure:

1. Size the gavage needle or red rubber catheter to the bird. The tube must be long enough to reach the crop at the base of the neck near the thoracic inlet. It is helpful to mark the location on the tube to indicate the depth at which it will have reached the crop. Lubricate the outside of the tube prior to placement.
2. Typically two people are required for the procedure, one person to restrain the bird and the other to perform the gavage.
3. Restrain the bird in an upright position and open the bird's mouth.

Fig. 5.11 Placement of a red rubber catheter for gavage or crop lavage in a chicken. Note the tape which marks the depth at which the tube will be in the crop.

4. Direct the gavage needle or tube into the left side of the mouth, over the tongue, and back toward the esophageal opening on right side of the mouth. The tracheal opening at the glottis is easy to identify on the floor of the mouth in birds and must be avoided.

5. Advance the tube down the esophagus into the crop. The holder should feel the tube as it passes down the right side of the neck.

6. Confirm placement by palpating the trachea separate from the tube, and the ball of the gavage needle or distal end of the flexible tube within the crop.

7. Occlude the esophagus and administer the compound. As a general guideline, the delivered volume should not exceed 10 mL/kg (Morton et al., 2001).

8. Continue occluding the esophagus as the tube is removed.

9. Keep the bird upright and monitor for any signs of respiratory distress.

grooming and general maintenance

- *Wing clipping*: A wing clip involves trimming a portion of the bird's primary wing feathers to prevent full flight. The procedure is used to prevent injury from flying, to prevent escape, or to assist in training (Schulte and Rupley, 2004). Although wing trims are common in pet birds, it is seldom indicated in a research environment. Avian pet medicine resources should be consulted for a description of the various techniques.

- *Toenail trimming*: Toenails are trimmed in small birds with nail clippers or suture removal scissors. A hand-held motorized grinding tool such as a Dremel® is typically used for larger birds. As in other species, such as dogs and cats, birds have a vessel running to the length of the nail that must be avoided. The vessel is readily visualized in birds with light-colored nails but difficult to discern in dark nails and can be difficult to avoid. Inadvertently cutting the quick is painful and will cause bleeding which can be substantial in larger birds. Kwik-Stop powder or silver nitrate sticks can be applied to stop bleeding.

- *Beak trimming*: Overgrown or sharp beaks are trimmed using nail clippers to trim the tip in small species such as quail or a hand-held motorized grinding tool such as a Dremel® for larger birds (Huss et al., 2008). Overzealous trimming may result in anorexia or bleeding.

- *Agricultural toe and beak trimming*: These procedures are sometimes used in the agricultural setting to reduce the incidence of trauma from aggressive pecking or mating (FASS, 2010). Toe trimming involves removal of the third phalanx of certain toes, whereas beak trimming involves partial removal of the upper and lower beak using a hot blade beak trimmer. Both procedures must be conducted in birds shortly after birth (FASS, 2010). These procedures remain controversial and are not typically used in a biomedical research setting. More information about these procedures can be found in the *Guide for the Care and Use of Agricultural Animals in Research and Teaching* (FASS, 2010).

necropsy

A necropsy involves the systematic postmortem examination of the body to include nearly all internal and external anatomical structures. They are useful in determining the cause of death or illness or the effects of experimental manipulations. A complete documentation of experimental manipulations should be obtained prior to initiation of a necropsy as this information correlated with necropsy findings will be helpful in drawing accurate conclusions. Careful examination of lesions or abnormalities, collection of tissue for histopathologic analysis, and of specimens for laboratory testing (e.g., complete blood count [CBC], serum chemistry, culture, cytology, polymerase chain reaction [PCR], etc.) will all be part of the necropsy process. The presence of infectious diseases, parasites, toxicities, nutritional deficiencies, tumors, and degenerative diseases are most likely to be diagnosed via necropsy. Management deficiencies, such as overcrowding or environmental fluctuations, or experimental procedure deficiencies such as improper monitoring during anesthesia, are more likely to be identified through a thorough history.

A necropsy should be performed as soon as possible after death to minimize tissue autolysis. If the procedure cannot be done immediately, the carcass can be stored at 4°C. Never freeze carcasses; this will cause significant postmortem changes and render histopathology largely useless.

Proper personal protective equipment (PPE) (at a minimum, gloves and a lab coat) is essential to minimize personnel exposure to zoonotic agents and other hazards. Potential zoonotic diseases in avian species include salmonellosis, chlamydiosis, mycobacteriosis, and campylobacteriosis. Additional PPE may be required if working with specific infectious agents. Always work in a well-ventilated area, preferably in a dedicated necropsy room with a downdraft table. If a zoonotic agent is suspected, the procedure should be performed in a laminar flow hood or biosafety cabinet. Most tissues are preserved in 10% buffered formalin, which is a known respiratory irritant and carcinogen. Nitrile gloves should be used when working with this agent, and a fumehood should be used when filling specimen containers.

When using formalin, a 10:1 fixative to tissue volume ratio is recommended for adequate tissue penetration. Trimmed tissues should be no more than 0.25 inch thick. Hollow organs (intestines, cloaca, trachea, etc.), except those from very small birds, are generally opened prior to fixation. Care must be taken not to touch the mucosa with instruments. Small cuts can be made in the intestine of smaller birds to allow formalin penetration. Very small pieces of tissue can be placed inside a labeled tissue cassette prior to placing in the formalin. Large organs (liver, kidney, heart, brain, and spleen) should be cut in half or in pieces to allow adequate penetration of the fixative. Generally at least 24 h is required for fixation.

Recommended Necropsy Equipment (Figure 5.12):

- A dedicated set of necropsy instruments that will not be used on live animals is preferred. Instruments include scalpel blades and handle, toothed forceps, hemostat, bone or poultry shears, and large and small dissecting scissors. Very fine instruments are required for small birds.
- Syringes and needles for blood or fluid collection, glass slides for impression smears, blood collection tubes and freezer-proof tubes, culture swabs, and a Bunsen burner for flaming/sterilizing instruments should be available for laboratory specimens collection.
- A dissection board. Push pins can be used in small birds to secure the extremities to the board.
- Specimen containers filled with 10% buffered formalin.
- Tissue cassettes for small organs or tissue sections.

Fig. 5.12 Necropsy setup for a small passerine. Note that the skin has been peeled from the mandible to the vent exposing the crop, keel, and coelom.

- Labels, a notepad for recording findings, indelible black marker or pencil.
- Scale, ruler, and camera.

Necropsy procedure:

- Although there are several ways to perform a necropsy, maintain a consistent approach to ensure that all systems are evaluated thoroughly and systematically.
- If the bird is still alive at presentation, observe the animal for abnormalities in general behavior and appearance. If abnormalities are observed, such as respiratory discharge, this can help in guiding the prosector in specimen collection at necropsy.
- After euthanasia, immediately collect blood samples before coagulation occurs, obtain a body weight, and perform an external exam. Evaluate the following:
  - Integument including plumage, beak, feet, and nails: examine for deformities, scales, scabs, bumps, swellings, and feather quality.
  - *Bones*: palpate for deformities or fractures.
  - *Cloaca*: examine for soiling, protrusions, and consistency of droppings.

- *Eyes*: examine for discharge, masses, scabs, and clarity.
- *Nares*: examine for discharge, masses.
- *Mouth*: examine for ulcers, plaques, cuts, masses, or foreign material.
- Wet the feathers with soapy water to avoid feathers from flying during the necropsy.
- Make a skin incision on the ventral midline from the mandible to the vent. Peel back the skin to expose the esophagus, trachea, crop, keel, and coelom.
- Examine the keel for bruising, pale areas, or weight loss. The breast muscle provides a good indication of general body condition. An overly rounded keel indicates obesity, whereas a sharply sloped keel indicates emaciation.
- Use large scissors or shears to cut the lateral pectoral muscles and ribs at the costrochondral junctions. At the cranial end of the breast (e.g., thoracic inlet), cut the corocoid and clavicle bones with shears. These bones are thick and may be difficult to cut in larger birds.
- Pull the keel up and away from the body and examine the air sacs. Air sacs should be translucent; opaque membranes, plaques, or fluid are abnormal and should be cultured or submitted for cytology. Remove the keel.
- Make a midline incision through the "abdominal" muscles. All major organs should now be visible.
- Examine all organs *in situ*. Normal organ appearance is as follows:
  - *Thyroids*: small, round, pink-red organs near the thoracic inlet along the carotid arteries.
  - *Liver*: homogenously dark brown with smooth, sharp edges. The right lobe is usually larger in psittacines.
  - *Spleen*: smooth, brownish-red, bean-shaped (passerines), round (psittacines and galliformes) or long, narrow (gulls) organ found between the ventriculus and proventriculus.
  - *Heart*: smooth, red with four chambers. Fat may be visible in the coronary groove.
  - *Lungs*: homogenously pink.
  - *Kidneys*: viewed by pushing overlying organs aside. Kidneys are red-brown and lobulated.

- *Proventriculus*: may be distended if filled with food.

- *Ventriculus*: firm muscular organ in birds that eat a seed or omnivorous diet. Thinner-walled and blends with the proventriculus in birds that eat fish or meat.

- *Gonads*: viewed by pushing overlying organs aside. Gonads sit at the anterior end of the kidney and can be used to determine sex of adult birds. Ovaries appear as a cluster of grapes while the testicle is round or bean-shaped.

- Cultures and impression smears of organs, if required, are best obtained as soon as possible to avoid contamination from other parts of the body. To obtain an organ culture swab, such as from liver, heat a blade over a Bunsen burner and lightly touch the heated blade to the organ surface. Make a cut with a new sterile blade through the sterilized organ surface. This creates a pocket in which the swab can be placed.

- Once the viscera have been examined, remove organs for closer examination and fixation. Tissues for microbiological or toxicologic analysis can be stored at –20°C.

- Cut along the base of the tongue through the posterior pharynx and hyoid apparatus and retract the tongue, esophagus, trachea, and thyroids. With continued traction, carefully dissect the lungs away from the ribs. Continue dissecting while applying traction all the way to the anterior pole of the kidneys.

- Cut around the vent and retract anteriorly to the kidneys, freeing the tissue with careful dissection.

- Remove the entire viscera.

- The kidneys lay against the pelvis and may not be able to be removed intact in very small birds. Kidneys can be removed with a section of the pelvis and fixed as a single block.

- Separate the heart, lungs, and crop from the posterior organs but cutting the esophagus and vessels just anterior to the liver.

- Open the esophagus. The mucosa should be smooth and tan or pink.

- The lungs can be sliced in segments like a loaf of bread, or can be prefilled with formalin by inserting a formalin-filled syringe into the trachea, occluding the tracheal opening, and injecting formalin into the air spaces. Lungs should float when placed in formalin.

- The heart can be opened to examine the valves, or in small birds, can be fixed whole.
- The liver can be sliced similarly to the lungs and placed in formalin.
- Open the proventriculus and ventriculus. The proventriculus mucosa has a slightly opaque surface. The ventricular mucosa of seed-eating or omnivorous birds has a horny lining called kaolin.
- Identify and measure the spleen.
- Inspect the muscles and joints of the wings, legs, and feet. Normal joint surfaces are smooth and glistening.
- Collect the brain by peeling the skin away from the scalp and cutting the calvarium with scissors or ronguers circumferentially around the head. The brain can be gently lifted at the front end and removed by cutting the cranial nerves as the brain is lifted up and back. The brain can be fixed in small birds by removing the entire head and placing in formalin.

references

Alworth, L.C., and Kelly, L.M. 2014. Endotracheal intubation and oral gavage in the domestic chicken. *Laboratory Animals.* 43:349–350.

Cabanac, M., and Aizawa, S. 2000. Fever and tachypnea in a bird (*Gallus domesticus*) after simple handling. *Physiology & Behavior.* 69:541–545.

Campbell, T.W. 1994. Hematology. In *Avian Medicine: Principles and Application*, edited by B.W. Ritchie, G.J. Harrison, and L.R. Harrison. Lake Worth, FL: Wingers Publishing, Inc., pp. 176–198.

DellaVolpe, A., Schmidt, V., and Krautwald-Junghanns, M.-E. 2011. Attempted semen collection using the massage technique in blue-fronted Amazon parrots (*Amazona aestiva aestiva*). *Journal of Avian Medicine and Surgery.* 25:1–7.

Divers, S.J. 2010. Avian diagnostic endoscopy. *Veterinary Clinics of North America: Exotic Animal Practice.* 13:187–202.

Fair, J., Paul, E., and Jones, J. Eds. 2010. *Guidelines to the Use of Wild Birds in Research*, Washington, D.C.: Ornithological Council. Online: http://www.nmnh.si.edu/BIRDNET/guide/index.html.

Federation of Animal Science Societies (FASS). 2010. *The Guide for the Care and Use of Agricultural Animals in Research and Teaching*, 3rd Edition. Campaign, IL: FASS. Online: http://www.fass.org.

Flammer, K. 1994. Antimicrobial therapy. In *Avian Medicine: Principles and Application*, edited by B.W. Ritchie, G.J. Harrison, and L.R. Harrison. Lake Worth, FL: Wingers Publishing, Inc., pp. 434–456.

Fox, R.A., and Millam, J.R. 2004. The effect of early environment on neophobia in orange-winged Amazon parrots (*Amazona amazonica*). *Applied Animal Behaviour Science*. 89:117–129.

Gee, G.F., Bertschinger, H., Donoghue, A.M., Blanco, J., and Soley, J. 2004. Reproduction in nondomestic birds: Physiology, semen collection, artificial insemination and cryopreservation. *Avian and Poultry Biology Reviews*. 15:47–101.

Geenacre, C.B., and Lusby, A.L. 2004. Physiologic responses of Amazon parrots (*Amazona* species) to manual restraint. *Journal of Avian Medicine and Surgery*. 18:19–22.

Halsema, W.B., Alberts, H., de Bruije, J.J., Luminei, J.T. 1988. Collection and analysis of urine from racing pigeons (*Columba livia domestica*). *Avian Pathology*. 17:221–225.

Harvey-Clark, C. 1990. Clinical and research used of implantable vascular access ports in avian species. *Proceedings of the Association of Avian Veterinarians*, Phoenix, AZ, 191–207.

Hawkins, P., Morton, D.B., Cameron, D., Cuthill, I.C., Francis, R., Freirc, R., ... Townsend, P. 2001. Laboratory birds: Refinements in husbandry and procedures. Fifth report of BVAAWF/FRAME/ RSPCA/UFAW joint working group on refinement. *Laboratory Animals*. 35:1–163.

Hughcs, B.O., and Black, A.J. 1976. The influence of handling on egg production, egg shell quality and avoidance behavior in hens. *British Poultry Science*. 17:135–144.

Huss, D., Poynter, G., and Langsford, R. 2008. Japanese quail (*Coturnix japonica*) as a laboratory animal model. *Laboratory Animals*. 37:513–519.

Influenza of non-domestic species: Supplemental training material. Avian influenza testing procedures. Online: http://www.zooani malhealthnetwork.org/FADSupplementalInfoInv04.pdf.

Kalmar, I.D., Moons, C.P.H., Meers, L.L., and Janssens, G.P.J. 2007. Psittacine birds as laboratory animals: Refinements and assessment of welfare. *JAALAS.* 46:8–15.

Lennox, A.M. 2006. Common procedures in other avian species. *Veterinary Clinics of North America: Exotic Animal Practice.* 9:303–319.

Morton, D.B., Jennings, M., Buckwell, A., Ewbank, R., Godfrey, C., Holgate, B., Inglis, I. et al. 2001. Refining procedures for the administration of substances. *Laboratory Animals.* 35:1–4

Quesenberry, K.E., and Hillyer, E.V. 1994. Supportive care and emergency therapy. In *Avian Medicine: Principles and Application*, edited by B.W. Ritchie, G.J. Harrison, and L.R. Harrison. Lake Worth, FL: Wingers Publishing, Inc., pp. 382–416.

Schulte, M.S., and Rupley, A.E. 2004. Avian care and husbandry. *Veterinary Clinics of North America: Exotic Animal Practice.* 7:315–350.

Senthilkumaran, C., Peterson, M., Taylor, M., and Bedecarrats, G. 2006. Use of a vascular access port for the measurement of pulsatile luteinizing hormone in old broiler breeders. *Poultry Science.* 85:1632–1640.

Wade, L. 2009. Restraint and administration of subcutaneous fluids and intramuscular injections in psittacine birds. *Laboratory Animals.* 38:292–293.

Zimmermann, N.G., and Dhillon, A.S. 1985. Blood sampling for the venous occipital sinus of birds. *Poultry Science.* 64:1859–1862.

6

resources

---

organizations

The American Association of Avian Pathologists (AAAP). Tel: 904-425-5735. Email: aaap@aaap.info. Website: www.aaap.info. AAAP provides resources related to managing and treating poultry. It publishes multiple manuals on these topics, including the *Avian Disease Manual, Avian Histopathology,* and others. AAAP also publishes a peer-reviewed journal, *Avian Diseases.* The Association has an annual meeting with proceedings available on its website.

The American Association for Laboratory Animal Science (AALAS). 9190 Crestwyn Hills Dr, Memphis, TN 38125-8538. Tel: 901-754-8620. Fax: 901-753-0046. Email: info@aalas.org. Website: www.acpv.info. AALAS represents and serves as a resource for all aspects of laboratory animal science, including veterinary and husbandry information. AALAS hosts an annual conference and publishes two peer-reviewed journals, *Comparative Medicine* and the *Journal of the American Association for Laboratory Animal Science.* The website includes a "Buyer's Guide" of vendors for animals and equipment for a variety of lab animal species.

The American College of Poultry Veterinarians (ACPV). 12627 San Jose Blvd, Suite 202, Jacksonville, FL 32223. Tel: 904-425-5735. Fax: 281-664-4744. Email: support@acpv.info. Website: http://www.acpv.info/. ACPV is the veterinary specialty board for poultry veterinarians recognized by the American Veterinary Medical Association. It conducts an annual conference with proceedings from limited years available to the public, and the most recent proceedings available to attendees.

The Association of Avian Veterinarians (AAV). PO Box 9, Teaneck, NJ 07666. Tel: 720-458-4111. Fax: 720-398-3496. Email: office@aav.org. Website: www.aav.org. AAV members are primarily veterinarians for companion animal birds, although it includes many practitioners who work with wildlife and other avian species. AAV publishes a peer-reviewed journal, the *Journal of Avian Medicine and Surgery* and holds an annual conference.

The Association for Assessment and Accreditation of Laboratory Animal Care (AAALAC) International. 5283 Corporate Dr., Suite 203, Frederick, MD 21703. Tel: 301-696-9626. Fax: 301-696-9627. Email: accredit@aaalac.org. Website: www.aaalac.org. AAALAC is a private organization that accredits laboratory animal programs to ensure high quality, humane animal care. Accreditation is voluntary. The website includes the AAALAC position statements, frequently asked questions, webinars, podcasts, and conference presentations. These resources primarily address AAALAC's interpretation and application of the guidance documents used by AAALAC to set accreditation standards.

The Institute for Laboratory Animal Research (ILAR). 500 5th St. NW, Washington, DC 20001. Tel: 202-334-2500. Email: dels@nas.edu. Website: http://dels.nas.edu/ilar/. ILAR is a unit in the Division on Earth and Life Studies of the National Research Council, the mission of which is to provide independent advice to the federal government. ILAR advises on matters related to the responsible use of animals in research. ILAR publishes multiple reports related to this topic, several of which include information on birds, such as the *Guide for the Care and Use of Laboratory Animals, Recognition and Alleviation of Pain in Laboratory Animals* and *Guidelines for the Humane Transportation of Research Animals.* In addition, ILAR publishes a peer-reviewed journal, *The ILAR Journal,* that has included articles related to the use of birds in research.

The United States Fish & Wildlife Service (USFWS). 1849 C Street, NW, Washington, DC 20240. Tel: 1-800-344-9453. Email is available through the website. Website: http://www.fws.gov/. The USFWS website includes information on laws and regulations that are applicable to the use of wild birds. The website also provides handbooks with related information and applications for associated permits. These laws include the Migratory Bird Treaty Act, Wild Bird Conservation Act, and the Endangered Species Act.

diagnostic laboratories

The major providers of diagnostic services for traditional animal species (e.g., IDEXX Laboratories, ANTECH Diagnostics) also typically offer tests for avian species. This section does not provide a comprehensive list of diagnostic laboratories but offers a few laboratories that specialize in exotic animal species to supplement or replace the services of major commercial service providers should the clinician desire.

Avian and Exotic Animal Clin Path Labs. Wilmington, OH. Tel: 800-350-1122. Fax: 937-383-3667. Website: http://www.avianexoticlab.com/. This lab offers standard wellness bloodwork, histopathology, bacteriology, and tissue testing for the metals most common in cases of toxicity.

Charles River Laboratories International, Inc. Tel: 800-772-3271. Website: http://www.criver.com/products-services/avian-vaccine-services. Charles River offers specific pathogen-free (SPF) chickens, SPF eggs, and infectious disease testing for pathogens primarily affecting poultry, chicken blood products, and other reagents as part of its Avian Vaccine Services.

The University of Georgia Infectious Disease Laboratory. 110 Riverbend Rd. Riverbend North, Room 150, University of Georgia, Athens, GA 30602. Tel: 706-542-8092. Fax: 706-583-0843. Website: http://www.vet.uga.edu/idl/. This laboratory offers full necropsy, histopathology, bacteriology, and polymerase chain reaction (PCR) sex determination in addition to specific infectious disease testing. The list of avian infectious disease tests available is extensive and includes tests for *Chlamydophila, Atoxoplasma,* and circoviruses.

The University of Georgia Poultry Diagnostic &. Research Center. 953 College Station Road, University of Georgia, Athens, GA 30602. Tel: 706-542-1904. Fax: 706-542-5630. Website: http://vet.uga.edu/avian/. This laboratory offers a variety of diagnostic tests specific to poultry, as well as necropsy and consultation services for clinicians.

The University of Miami Avian and Wildlife Laboratory. 1600 NW 10h Ave, RMSB 7101A, Miami, FL 33136. Tel: 800-596-7390. Fax: 305-243-5662. Website: http://cpl.med.miami.edu/avian-and-wildlife. This laboratory offers standard wellness bloodwork including electrophoresis, as well as PCR sex determination and limited infectious disease testing.

Michigan State University Diagnostic Center for Population and Animal Health. Tel: 517-353-1683. Fax: 517-353-5096. Website: www.

animalhealth.msu.edu. 4125 Beaumont Rd., Lansing, MI 48910. The Diagnostic Center for Population and Animal Health (DCPAH) is a full-service veterinary diagnostic laboratory offering more than 800 tests in 11 service sections for all species. Tests available for birds range from complete blood count (CBC) and serum chemistry profiles to *Chlamydophyla* PCR and avian influenza assays.

Louisiana Animal Disease Diagnostic Laboratory. LSU, River Road. Room 1043, Baton Rouge, LA 70803. Tel: 225-578-9777. Fax: 225-578-9784. Website: http://laddl.lsu.edu. This laboratory offers testing for the metals most commonly associated with toxicity, including lead and zinc. The website provides details on proper preparation and collection of samples for these tests as well.

Zoologix. 9811 Owensmouth Ave, Suite 4, Chatsworth, CA 91311. Tel: 818-717-8880. Email: info@zoologix.com. Website: http://www.zoologix.com/avian/index.htm. This laboratory offers PCR tests for many avian infectious diseases.

publications

Books

General avian biology and medicine

Avian Medicine. By J. Samour. 2008. ISBN 9780723434016. 540 pages. 2nd edition. Elsevier Mosby, New York, NY.

Avian Medicine: Principles and Application. Edited by L.R. Harrison. 1999. ISBN 9780967406602. 1384 pages. HBD International, Delray Beach, FL. Note: This title is also available under ISBN 0963699601.

Avian Medicine and Surgery in Clinical Practice: Companion and Aviary Birds. By Bob Doneley. 2010. ISBN 9781840765922. 336 pages. CRC Press Taylor & Francis Group, LLC, Boca Raton, FL.

Clinical Avian Medicine and Surgery. By G.J. Harrison and T. Lightfoot. 2006. ISBN 9780975499405. 1008 pages. Spix Publishing, Palm Beach, FL.

Essentials of Avian Medicine and Surgery. By Brian Coles. 2008. ISBN 9780470691564. 392 pages. 3rd edition. Blackwell Publishing Ltd., Oxford, UK.

Handbook of Avian Medicine. By T.N. Tulley, G.M. Dorrestein, and A.K. Jones. 2009. ISBN 9780702028748. 478 pages. 2nd edition. Elsevier Saunders, Philadelphia, PA.

Manual of Avian Practice. By A.E. Rupley. 1997. ISBN 9780721640839. 556 pages. Elsevier Saunders, Philadelphia, PA.

Avian medicine-poultry focus

Avian Disease Manual. Edited by M. Boulianne. 2013. ISBN 9780978916343. 300 pages. 7th edition. American Association of Avian Pathologists, Jacksonville, FL.

Backyard Poultry Medicine and Surgery. By C.B. Greenacre and T.Y. Morishita. 2014. ISBN 9781118403846. 368 pages. John Wiley & Sons, Ames, IA.

BSAVA Manual of Farm Pets. By V. Roberts and F. Scott-Park. 2008. ISBN 9781905319039. 320 pages. John Wiley & Sons, Ames, IA.

Diseases of Poultry. Edited by D.E. Swayne. 2013. ISBN 9780470958995. 1408 pages. 13th edition. Wiley-Blackwell, Ames, IA.

Avian medicine-pigeon focus

BSAVA Manual of Raptors, Pigeons and Passerine Birds. By J. Chitty and M. Lierz. 2008. ISBN 9781905319046. 352 pages. John Wiley & Sons, Ames, IA.

The Flying Vet's Pigeon Health and Management. By C. Walker. 2000. ISBN 9781876677916. 322 pages. Knox Veterinary Clinic, Knox, IN.

Pigeon Health and Disease. By D.C. Tudor. 1991. ISBN 9780813812441. 244 pages. Iowa State University Press, Iowa City, IA.

A Veterinary Approach to Pigeon Health. By D. Marx. 1997. ASIN B0006R6DPI. 212 pages. Racing Pigeon Digest Pub, Emigrant, MT.

Avian medicine-wild bird focus

BSAVA Manual of Raptors, Pigeons and Passerine Birds. By J. Chitty and M. Lierz. 2008. ISBN 9781905319046. 352 pages. John Wiley & Sons, Ames, IA.

Diseases of Wild Waterfowl. By G.A. Wobeser. 2012. ISBN 9781461559511. 324 pages. 2nd edition. Springer Science & Business Media, New York, NY.

Fowler's Zoo and Wild Animal Medicine. By R.E. Miller and M.E. Fowler. 2014. ISBN 9781455773992. 792 pages. 8th edition. Elsevier Saunders, St.Louis, MO.

Zoo and Wild Animal Medicine Current Therapy. By M.E. Fowler and R.E. Miller. 2007. ISBN 9781416064633. 512 pages. 6th edition. Elsevier Saunders, St. Louis, MO.

Avian clinical pathology

Avian and Exotic Animal Hematology and Cytology. By T.W. Campbell and C.K. Ellis. 2013. ISBN 9781118710661. 287 pages. 3rd edition. John Wiley & Sons, Ames, IA.

Exotic Animal Hematology and Cytology. Edited by T.W. Campbell. 2015. ISBN 9780813818115. 424 pages. 4th edition. Wiley-Blackwell, Ames, IA.

Laboratory Medicine: Avian and Exotic Pets. By A.M. Fudge. 2000. ISBN 9780721676791. 486 pages. Saunders, Philadelphia, PA.

Laboratory birds

Anesthesia and Analgesia in Laboratory Animals. By R. Fish, P.J. Danneman, M. Brown, and A. Karas. 2011. ISBN 9780080559834. 672 pages. 2nd edition. Academic Press, London, England.

Comfortable Quarters for Laboratory Animals. Edited by V. Reinhardt and A. Reinhardt. 2002. 106 pages. 9th edition. Animal Welfare Institute, Washington, DC. This book has a chapter on laboratory chickens that is also available online at http://awionline.org/pubs/cq02/Cq-chick.html.

Guide for the Care and Use of Laboratory Animals. Institute for Laboratory Animal Research. 2011. ISBN 9780309186636. 248 pages. 8th edition. National Academy Press, Washington, DC.

The Guide for the Care and Use of Agricultural Animals in Research and Teaching. Federation of Animal Science Societies. 2010. ISBN 9781884706110. 177 pages. 3rd edition. Federation of Animal Sciences, Champaign, IL.

Guidelines for the Humane Transportation of Research Animals. Institute for Laboratory Animal Research. 2006. ISBN 9780309164771. 164 pages. National Academy Press, Washington, DC.

Laboratory Animal Medicine. Edited by L.C. Anderson, G. Otto, K.R. Pritchett-Corning, M.T. Whary, and J.G. Fox. 2015. ISBN 9780124095274. 1746 pages. 3rd edition. Academic Press, New York, NY.

Recognition and Alleviation of Pain in Laboratory Animals. Institute for Laboratory Animal Research. 2009. ISBN 9780309149112. 196 pages. National Academy Press, Washington, DC.

The UFAW Handbook on the Care and Management of Laboratory and Other Research Animals. Edited by R.C. Hubrecht and J. Kirkwood. 2010. ISBN 9781444318784. 848 pages. 8th edition. John Wiley & Sons, Chichester, UK.

The Zebra Finch: A Synthesis of Field and Laboratory Studies. By R.A. Zann. 1996. ISBN 9780198540793. 335 pages. Oxford University Press, Oxford, UK.

Formularies

Essentials of Avian Medicine and Surgery. By Brian Coles. 2008. ISBN 9780470691564. 392 pages. 3rd edition. Blackwell Publishing Ltd., Oxford, UK.

Exotic Animal Formulary. By J.W. Carpenter. 2013. ISBN 9781437722642. 744 pages. 4th edition. Elsevier Saunders, St. Louis, MO.

Periodicals

Avian Diseases is the peer-reviewed, quarterly journal published by the American Association of Avian Pathologists. ISSN 0005-2086.

Comparative Medicine is a peer-reviewed, bimonthly publication of the American Association for Laboratory Animal Science. ISSN 1532-0820.

ILAR Journal is the peer-reviewed, monthly journal published by the Institute for Laboratory Animal Research. Online ISSN 1930-6180. Print ISSN 1084-2020.

Journal of the American Association for Laboratory Animal Science (JAALAS) is a peer-reviewed, bimonthly publication of the American Association for Laboratory Animal Science. ISSN 1559-6109.

Journal of Avian Medicine and Surgery is the peer-reviewed, quarterly journal published by the Association of Avian Veterinarians. Online ISSN 1938-2871. Print ISSN 1082-6742.

Journal of Exotic Pet Medicine is a quarterly journal published by the Association of Exotic Mammal Veterinarians (AEMV) but includes articles on birds and other nonmammalian companion animal species. ISSN 15575063.

Journal of Zoo and Wildlife Medicine is the peer-reviewed, quarterly publication of the American Association of Zoo Veterinarians (AAZV) and the European Association of Zoo and Wildlife Veterinarians (EAZWV). Online ISSN 1937-2825. Print ISSN 1042-7260.

Lab Animal Magazine is a peer-reviewed journal that also publishes a *Lab Animal's Buyers' Guide* with vendors for many supplies, both as a hard copy and in an online, searchable format. ISSN 0093-7355.

Laboratory Animals is a peer-reviewed journal that covers a variety of topics, including the management and use of birds in research. Online ISSN 1758-1117. Print ISSN 0023-6772.

Veterinary Clinics of North America: Exotic Animal Practice is published three times a year, and each issue focuses on a specific topic related to exotic companion animal medicine. ISSN 1094-9194.

Electronic Publications and Websites

Field Manual of Wildlife Diseases—General Field Procedures and Diseases of Birds. Published by the U.S. Geological Survey of the U.S. Department of the Interior is available at http://www.nwhc.usgs.gov/publications/field_manual/. This reference includes both toxins and infectious agents that pose a risk to wild birds.

Guidelines to the Use of Wild Birds in Research (2010). Third edition of the publication by the Ornithological Council can be found at http://www.nmnh.si.edu/BIRDNET/guide/index.html. This document provides an overview of regulatory and practical considerations when working with birds in the field.

Laboratory Bird: Refinements in Husbandry and Procedures (2001). Fifth report of the BVAAWF/FRAME/RSPCA/UFAW Joint Working Group on Refinement, edited by P. Hawkins, is available on the website of the Royal Society for the Prevention of Cruelty to Animals (RSPCA) http://science.rspca.org.uk/sciencegroup/researchanimals/implementing3rs/refinement, as originally published in *Laboratory Animals.* The RSPCA also publishes documents entitled "Good practice for housing and care" for domestic fowl, ducks and geese, pigeons, quail and zebra finches, which can be found by searching from the RSPCA home page (http://www.rspca.org.uk/home). These documents have limited references, but may provide some information useful to housing these animals in a research environment.

---

vendors

Sources of Animals

Most avian species, with the exception of chickens, are not readily available from commercial vendors. Local bird breeders can be often be located by contacting bird clubs in the area that focus on the species of interest (e.g., racing pigeon clubs and exotic pet birds). Breeders of birds for pets and show animals often do not maintain the same level of quality control or disease monitoring as many laboratory animal programs have come to expect from rodent vendors, and each will need to be evaluated on an individual basis and accepted with quarantine and entry procedures based on a risk assessment for the given institution. Although many bird breeders have less than ideal health programs, they are likely preferable to obtaining birds directly from a pet store where animals from multiple sources can be mixed. Birds are easily stressed by transport, so the shortest shipping distance should be considered when possible. When evaluating a vendor and developing a quarantine program, realize that ectoparasites are likely the most common pathogen, with endoparasites and zoonotic diseases, such as *Chlamydophila*, also posing some risk. Specific animal vendors, other than that listed hereunder, are not mentioned as institutions will need to evaluate vendors for compatibility with their research programs.

Charles River Laboratories International, Inc. Tel: 800-772-3271. Website: http://www.crivcr.com/products-services/avian-vaccine-services. This vendor provides specific pathogen free chickens, eggs, and reagents.

Feed and Nutritional Supplements

Kaytee (http://www.kaytee.com/) offers diets targeted toward psittacines and passerines typically kept as pets, including finch and canary diets. It also offers a variety of enrichment and housing products.

Lafeber (http://lafeber.com/pet-birds/shop/) offers a variety of diet options, primarily for psittacines. These include critical care diets (Emeraid products) and neonate formulas.

Mazuri, PMI Nutrition International (http://www.mazuri.com/) offers formulated diets for waterfowl, psittacines, passerines, and other traditional zoo species. The diets include options for maintenance, breeding, and formulas for neonates.

Purina LabDiet provides diets suitable for chicks, laying poultry, passerines, and large and small psittacines. Additional supplements may be required for breeding birds other than chickens. Information available at: http://www.labdiet.com/Products/StandardDiets/Poultry/index.htm.

Zeigler (http://www.zeiglerfeed.com/html/). Products include adult maintenance and breeder diets for psitttacines and passerines, as well as neonate formulas.

Zupreem (http://www.zupreem.com/). Products include adult maintenance and breeder diets for psitttacines, as well as neonate formulas.

Caging, Veterinary, and Miscellaneous Equipment

Possible Sources of Equipment and Supplies

Item	Sources
Identification	3, 5, 11, 12, 13
Anesthesia and monitoring	4, 7, 8, 14, 15, 17
Surgery	6, 7, 14, 16, 17
Miscellaneous veterinary supplies	7, 8, 17
Caging and husbandry supplies	1, 2, 9, 10, 16

Contact information for vendors

1. Allentown. Tel: 800-762-2243. Website: http://www.allentowninc.com/en/large-animal-housing/poultry/.

2. Alternative Design. 3055 Cheri Whitlock Drive. Siloam Springs, AR 72761. Tel: 800-320-2459. Website: http://www.altdesign.com/.

3. Avid Identification Systems, Inc. 3185 Hamner Ave., Norco, CA 92860. Tel: 800-336-2843.

4. BASi Vetronics. 2701 Kent Ave., West Lafayette, IN 47906. Tel: 800-845-4246. Website: http://www.basinc.com/products/vetronics/products/smvent.php.

5. Bird Banding Laboratory of the U.S. Geological Survey. Website: https://www.pwrc.usgs.gov/BBL/manual/order.cfm.

6. Ellman. 400 Karin Ln., Hicksville, NY 11801. Tel: 800-835-5355. Website: http://www.ellman.com/index.html.

7. Jorgensen Labs. 1450 Van Buren Ave., Loveland, Co 80538. Tel: 800-525-5614.

8. Instech. Plymouth Meeting, PA, USA. Tel: 800-443-4227. Website: http://www.instechlabs.com/.

9. Lenderking Caging Products. 8370 Jumpers Hole Road, Millersville, MD 21108. Tel: 410-544-8795. Website: http://www.lenderking.com/avian.html.

10. LGL Animal Care Products, Inc. College Station, TX 77845. Tel: 979-690-3434. Website: http://www.lglacp.com/.

11. L & M Bird Leg Bands, Inc. PO Box 2636, San Bernardino, CA 92406. Tel: 909-882-4649. Website: http://www.lmbirdlegbands.com.

12. National Band & Tag Company. 721 York St, Newport, KY 41072. Tel: 859-261-2053. Website: http://nationalband.com/poultry-bands/.

13. National Finch and Softbill Society. 918 Georgia Ave, Etowah, TN 37331. Tel: 509-724-0107. Website: http://nfss.org/shop/bands/.

14. SAI Infusion Technologies. Tel: 847-356-0321. Website: http://www.sai-infusion.com/.

15. Smiths Medical. 5200 Upper Metro Place, Suite 200, Dublin, OH 43017, U.S.A. Tel: 800 258 5361. Website: http://www.smiths-medical.com/.

16. Suburban Surgical Co., Inc., 275 Twelfth Street. Wheeling, IL 60090. Tel: 800-323-7366. Website: http://www.suburbansurgical.com/.

17. Veterinary Specialty Products. 10504 W 79th St. Shawnee, KS 66214. Tel: 800-362-8138. Website: http://www.vetspecialtyproducts.com/index.cfm?fuseaction=home.main.

Index